Praise for *Deadliest Sea*

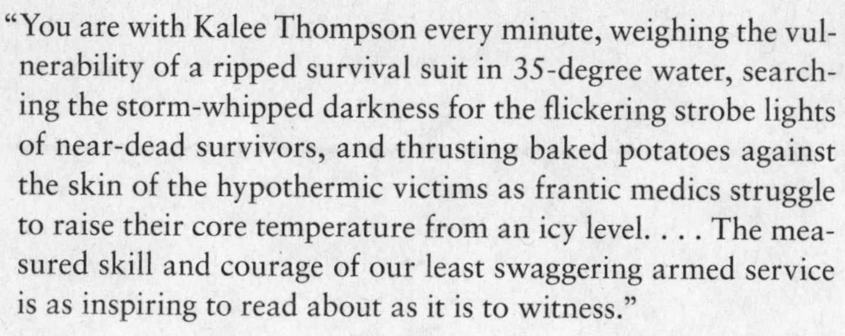
"You are with Kalee Thompson every minute, weighing the vulnerability of a ripped survival suit in 35-degree water, searching the storm-whipped darkness for the flickering strobe lights of near-dead survivors, and thrusting baked potatoes against the skin of the hypothermic victims as frantic medics struggle to raise their core temperature from an icy level. . . . The measured skill and courage of our least swaggering armed service is as inspiring to read about as it is to witness."

—*Washington Post*

"Like *Into Thin Air* and *The Perfect Storm*, *Deadliest Sea* is a gripping story of death and survival in one of the world's most dangerous places. It is also a portrait of heroism. . . . We need more stories like this." —*Popular Mechanics*

"A great story. . . . Kalee Thompson reports on every detail of the disaster and subsequent rescue. It's like an episode of *The Deadliest Catch,* when one of the ships goes down and, out of nowhere, a heroic team of Coast Guard specialists swoops in to save as many souls as they can. Needless to say, it's riveting."

—GQ.com, *The Verge* blog

"Books about disaster in the Alaskan fishing industry are common, but titles as good as this one are rare. . . . This extraordinary book is accessible to every level of interest in those who go out to the sea in ships and may well keep most of them up all night."

—*Booklist*

"Thompson's heavy-duty reporting in *Deadliest Sea* offers a depth of background and information that would be impossible to reproduce in a video or television format."

—Boats.com

"Thoroughly researched, *Deadliest Sea* provides great insight into the challenges that face both rescuer and survivor alike."

—*Boating Magazine*

"One of the greatest rescue stories in maritime history."

—*Laconia Citizen* (New Hampshire)

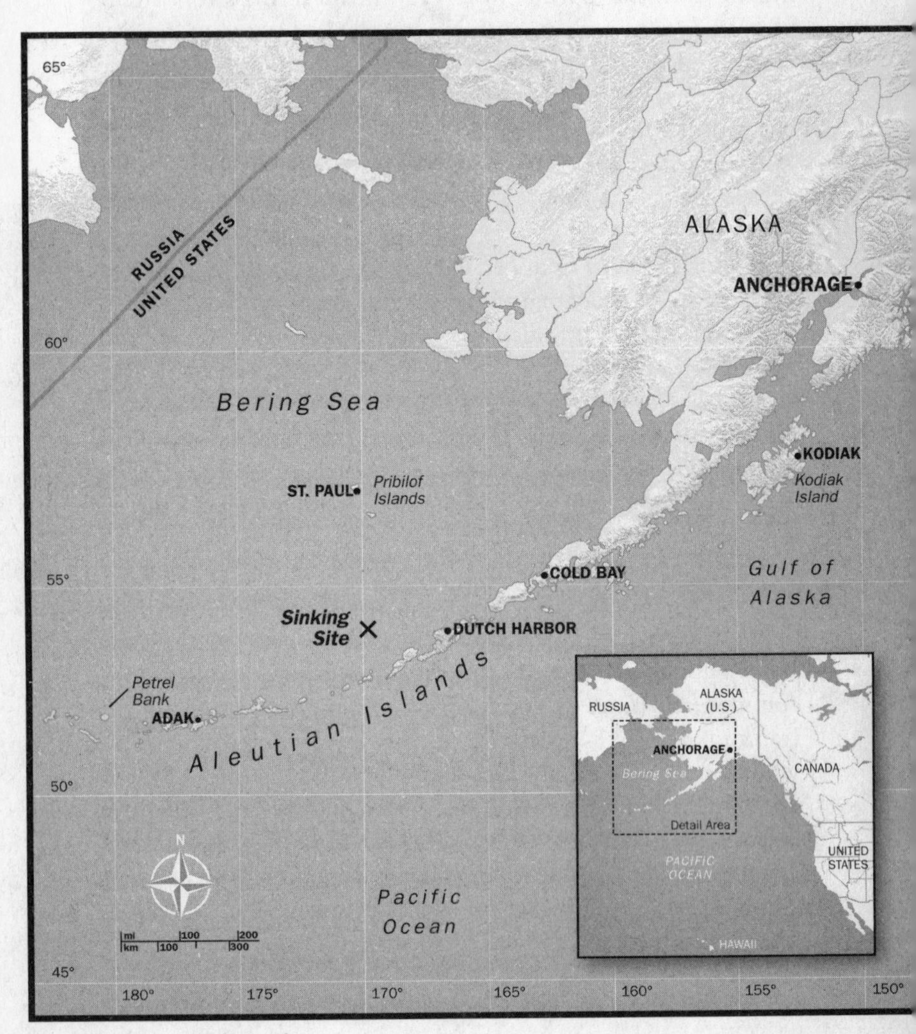

Map by Steve Walkowiak

DEADLIEST SEA

The Untold Story Behind the Greatest Rescue in Coast Guard History

KALEE THOMPSON

HARPER

NEW YORK . LONDON . TORONTO . SYDNEY

HARPER

A hardcover edition of this book was published in 2010 by William Morrow, an imprint of HarperCollins Publishers.

Portions of the text previously appeared in slightly different form in the article "Ranger Down" in the July 2008 issue of *Popular Mechanics*.

FIRST HARPER PAPERBACK PUBLISHED 2011.

Designed by Jamie Kerner

Library of Congress Cataloging-in-Publication Data is available upon request.

ISBN 978-0-06-176630-5

11 12 13 14 15 OV/RRD 10 9 8 7 6 5 4 3 2 1

This book is dedicated to all the fishermen who never came home.
And to the families who still miss them every day.

Contents

	Prologue	1
1	The *Alaska Ranger*	7
2	"Go to the Suits!"	31
3	Always Ready	48
4	Best Speed	68
5	"Abandon Ship!"	92
6	The Observers	113
7	Alone in the Waves	135
8	Swimmer in the Water	157
9	Sick at Sea	178
10	Man Down	195
11	Out of the Cold	215
12	Death at the Extremes	231
13	The Final Search	249
14	The Investigation	265
	Epilogue	286
	Notes	291
	Acknowledgments	307

Deadliest Sea

Prologue

"Mayday. Mayday. Mayday. This is the *Alaska Ranger.*"

The words cut through the constant buzz of static that filled the bare, windowless cubicle on Kodiak Island where David Seidl was the watchstander on duty. He pressed his thumb into a black button on his microphone.

"Station calling, this is the United States Coast Guard, Kodiak, Alaska, Communications Station, over."

Seconds later, a response broke into the fuzz: "Yeah, United States Coast Guard, this is the *Alaska Ranger.* Our position is 5, 3, 5, 3.4—53, 53.4 north, 1, 6, 9, 5, 8.4—1, 6, 9, 5, 8.4 west. We are flooding, taking on water in our rudder room. We are flooding by the stern."

Seidl had been trained to look at the clock the moment a distress call came in. It was 10:46 Zulu. Just before 3:00 A.M. Alaska Standard Time. He knew the checklist of critical information: Name and description of vessel, location, nature of the emergency, and POB—the number of people on board.

"*Alaska Ranger,* this is COMMSTA Kodiak." Seidl spoke

evenly into the microphone. "Roger, good copy on position. Understand you are flooding, taking on water from the stern. Request to know number of persons on board, over."

More static, and the gravity of the situation became clear: "Number of persons is, uh, forty-seven people on board, okay?"

It had been light outside ten hours before when Seidl left his apartment to drive to work. He was twenty-six years old with closely cropped brown hair and wire-rimmed glasses, and he had been a watchstander at the communications station for just over a year. Seidl worked three or four twelve-hour shifts each week, periodically switching from days to nights. It was Saturday, March 22, 2008, and his shift ran from 5:30 p.m. to 5:30 a.m. The next day was Easter, but Seidl didn't have any particular plans. Sleep, then report to work again the next evening.

The Kodiak job was Seidl's first Coast Guard assignment. He'd become interested in the Coast Guard when the service became part of the Department of Homeland Security, after 9/11. Seidl had studied intel in college and thought that the Coast Guard would be a good place to use his degree. Veteran operations specialists had told him Alaska was a sweet assignment—always busy, lots of big cases. But once he got there, he figured they must have been talking about Juneau. Those guys obviously hadn't been in Kodiak. The area was known for world-class hunting and fishing, but those things weren't a draw for Seidl. He took a part-time job washing cars at Avis. It killed the time.

It was 34°F and breezy as Seidl passed Kodiak's twenty-five-bed hospital, the island's single high school, and the Gas N Go—one of just four public fuel stations in town. He slowed his Jeep through the "Y," Kodiak's busiest intersection, where the island's first traffic light was under construction. David could see the Kodiak fishing fleet down in the harbor, mostly small catcher boats that supply the fish processing plants in town. Kodiak is the third most profitable fishing port in the country, and the industry defines the place.

So does the Coast Guard. More than a thousand active duty men and women work at Air Station Kodiak. Add civilian employees and family members and the Coast Guard community numbers close to three thousand. While most single men and women serve only two years at the remote air station (the sprawling facility is a hand-me-down from the Navy, which ran a base on the island during World War II), married personnel are more likely to call Kodiak home for three, four, even five years. The odd Coastie falls in love with the place and finds ways to lengthen the assignment. Sometimes that means passing up promotions or quitting the service altogether.

About half of the Kodiak Coasties live on base, most in single-family homes in cookie-cutter neighborhoods that look just the same as many suburban communities in the Lower 48. The white, beige, and gray houses line up on curving streets with names like Albatross and Pigeon Point. Some of the largest sit at the end of culs-de-sac, with pickup trucks or minivans parked out front. Anchoring the winding lanes are playgrounds with jungle gyms and plastic slides and the iconic Alaskan swing: a bulbous orange fishing buoy hanging from a frayed line.

Bachelors and the most junior enlisted men and women live in barracks on the far side of the runway, closer to the base's movie theater, pizza pub, and bowling alley, Tsunami Lanes—named for the tidal wave that flattened Kodiak in 1964. The rebuilt town is far from charming, with its weathered, wooden structures and their shedding paint and faded signs, all huddled around the crowded boat harbor.

A stink grows as the road rises away from the bay. The canneries that line the waterfront lane known as Shelikof Street process dozens of types of fish. But the staple catch is walleye pollack, a shimmery silver fish that resembles an oversized sardine. In an hour's time, thousands of pollack can be sucked by vacuum hose from the belly of a small catcher boat, fed onto a conveyer belt and through a maze of stainless steel processing equipment, where they're skinned, boned, and pulped into a thick, dry dough that resembles a huge batch of mashed potatoes. Sugar and preservatives are added before the mash is shaped into blocks and frozen in trays. Later, the frozen pulp will be molded into fish sticks, fast-food fish sandwiches, imitation crab, and all the other products known to the American public simply as "fish." Few town residents ever see the inside of the fish plants. Instead, all they see is the row of corrugated metal buildings along the waterfront, the obese bald eagles that linger near the fish scraps in the canneries' Dumpsters, and the multinational workforce that keeps the plants running—Filipinos, Mexicans, Samoans, and native Alutiiqs from the tiny, isolated villages on the far side of Kodiak Island.

Seidl pulled into the paved lot just below Communications Station Kodiak (or COMMSTA, pronounced "com stay"). The station is one of just three remaining high-frequency (HF) communication sites run by the Coast Guard that remains manned full-time. The single-story, white building is surrounded by nearly thirty communication towers—a candy-cane-striped array that allows the station to pick up high- and medium-frequency radio communications from all over the world.

At work, some of the guys joked that only Ted Stevens ("Uncle Ted" as Alaskans call the six-term senator) kept the fifty-some COMMSTA workers employed. After all, the additional Coasties sent to run the station brought money and jobs into the local economy. But it was obvious to almost everyone that it was only a matter of time before this station, too, would be automated.

Until then, the watchstanders in Kodiak share a series of twenty-four-hour duties. Four times a day they transmit faxes from the National Weather Service to ships at sea, and every two hours they broadcast navigation warnings. COMMSTA keeps radio contact, or guard, with the Coast Guard aircraft based out of Kodiak. Every fifteen minutes, the station checks in with any helicopter in flight, and every thirty minutes it makes contact with any airborne Hercules C-130, the Coast Guard's primary fixed-wing aircraft. Finally, each watchstander spends four hours of each twelve-hour shift in the "distress room," a small carpeted office whose single desk is stacked with radios.

Seidl began his four-hour distress room shift at 1:30 A.M. on Sunday, March 23. In the thirteen months he had been standing duty in Kodiak, he had never heard any true emergency call coming through the speakers. The HF radio just emitted that constant, gnawing hum of static. The ships in Alaska seemed to know what they were doing. If they had a problem, they could often fix it themselves or call a nearby ship for help with a VHF radio or satellite phone. Usually the small boats were the ones that got in trouble—bad weather or whatever. Another boat could often deliver a pump or give them a tow. The troubled vessel might let the Coast Guard know what was going on, or maybe not.

There were several radios on the bank of electronics in front of Seidl, each set to a different frequency. One was tuned to 2182, the international hail and distress channel. Another was set to 4125, the international frequency for mariners. His job was to sit in the room and listen for a break in the static. He was used to nothing happening. For a year he'd sat in there and nothing had happened.

To kill time, he had devised a workout routine. Push-ups, sit-ups, and squats. For an hour, he cycled through the exercises, growing damp in his T-shirt and shorts. He'd only been finished for a couple minutes when he head the first "Mayday" cut through the static, over 2182.

"*Alaska Ranger*, THIS IS COMMSTA, ROGER," Seidl replied to the report of the crew's size. "Understand forty-seven persons on board. Request vessel description, over."

"We are a factory trawler," the ship's officer answered. "We're one hundred eighty-four feet in length, black hull, white trim, okay?"

"*Alaska Ranger,* this is COMMSTA Kodiak. Roger. Understand factory trawler, one, eight, four feet in length, black hull and white trim. Stand by one, over."

Ten seconds later, Seidl hailed the ship again, and repeated its reported latitude and longitude, a position north of Alaska's Aleutian Island Chain, 140 miles from the nearest port, in the middle of the Bering Sea.

"*Alaska Ranger,* this is COMMSTA. Confirm position, 5, 3, 5, 3.4 north, 1, 6, 9, 5, 8.4 west. Over."

"That's a roger, 53, 53.4 north, 1, 6, 9, 5, 8.4 west, okay?"

"*Alaska Ranger,* this is COMMSTA. Request to know if you are able to keep up with the flooding at this time, over?" Seidl asked at 2:49 A.M.

"Uh, negative," the voice came back. "Negative. . . . The fire pumps cannot keep up."

The ship was more than eight hundred miles away.

Chapter One

The *Alaska Ranger*

From the window of the tiny turboprop, Julio Morales stared down on jagged, snow-sheathed summits. He'd never seen such huge mountains, never been in such a small airplane. He'd been handed a pair of Styrofoam earplugs when he boarded the flight at the Anchorage airport and was told to keep his seat belt on. And then they'd taken off, up and out over Cook Inlet, the long bay that leads from Alaska's largest city toward the open ocean. They were headed west, over the towering volcanic peaks of the Alaskan Peninsula, and along the Aleutian Island Chain to the fishing port of Dutch Harbor. It was about as far west as you could go, farther west than Julio had ever been.

Julio was forty years old, but he looked much younger, with big, wet brown eyes and smooth, round cheeks. He had a couple of cousins who'd worked in Dutch Harbor, one at a fish processing plant, and the other on a boat. He liked their stories of Alaska. Just the idea of the place appealed to him—so big, so empty.

As the plane descended, Julio couldn't see anything: no city, no airport, no lights. Outside the window was just a wall of white. And then, all of a sudden, they were on the runway, a clipped stretch of asphalt laid across a narrow spit of land dividing two large bays. Julio climbed down a metal stepladder straight into the cold, late-winter afternoon. He picked up his green oversized army duffel inside the one-room airport and found the waiting van. It had been sent by his new employer, the Fishing Company of Alaska (FCA), to deliver him a mile down the road to the Grand Aleutian hotel.

If it were located anywhere else in the country, the Grand Aleutian's direct competition would be a typical Holiday Inn. The comforters are polyester, the bathroom floors are linoleum, and the tubs are small and scuffed. Even many of the nonsmoking rooms smell like stale cigarettes. The hotel is the nicest in Dutch Harbor. But it's also the only hotel in Dutch Harbor—with the exception of bunkhouses for processing-plant workers and government fishery employees. The three-story, 110-room, crescent-shaped hotel hugs Margaret Bay, a tiny inlet that attracts geese and ducks and the odd sea lion. In the summertime, when Dutch Harbor turns lush and green, a handful of extreme birders will book at the Grand Aleutian, where rooms start at $160 a night. There are the odd adventure travelers, planning long treks across the treeless islands, or kayaking expeditions in some of the roughest, coldest waters on the planet. There's some historical tourism from World War II veterans who were stationed in Dutch Harbor during the Pacific campaign, and, in recent years, visits from die-hard fans of the Discovery Channel show *Deadliest Catch*, which follows the crews of crab boats that work out of the port.

For the most part, though, Dutch Harbor is far off the wildlife-and-glaciers tourist circuit that draws more than a million visitors to Alaska each summer. Most of the guests at the Grand Aleutian are in town for business. They're men who own boats, who buy fish, or fix ships. And they're fishery workers waiting to get on a boat or waiting for a flight home—sometimes for days and days and days. The hulking hotel is run by UniSea, Inc., a Japanese-owned seafood company that operates the largest of five fish-processing plants in town. Officially, the entire populated area is part of the city of Unalaska. Many locals use that more accurate name, especially the native Aleuts who have inhabited the islands for thousands of years. To outsiders and most fishermen, the whole place is just "Dutch."

Despite a full-time population of just 4,300 people, Dutch Harbor/Unalaska is the top fishing port in the United States in terms of volume of catch. The town has held the title for twenty years running. In 2008, the two-hundred-odd fishing boats that sail out of Dutch hauled in 612.7 million pounds of fish—more than 13 percent of the total U.S. catch, and worth $195 million dollars.

Most of the full-timers live across the bridge from the airport and the Grand Aleutian hotel. "The Bridge to the Other Side," it's called. Several square blocks of wooden homes line the black-sand waterfront looking out into Iliuliuk Bay. Though many of the houses are brightly painted—blue, pink, and green—many are deep in disrepair, with rotting wood and broken windows and yards strewn with old fishing gear. A modern library, medical clinic, and high school seem at odds with the dilapidated houses that surround them. The town buildings were improved in the 1980s and 1990s, when the community was flush with fish money. Now Unalaska has an Olympic-size pool in its aquatic center and an indoor track and racquetball courts at the town recreation facility.

Julio didn't go to the gym or the pool. But he did explore around the UniSea plant, just down the road from the hotel. It was Rosel Garcia's idea to get outside. Julio didn't want to miss a call from the company. They'd been told they would be called when their boat arrived, and that they should wait at the hotel until then. Julio wasn't sure they should leave the building. But they'd been in town for a couple of days already, and they wanted to see more of the island. Julio had noticed Rosel in the airport in Anchorage, a black guy speaking Spanish on his cell phone. Julio could tell from his accent that Rosel was from Central America—Honduras, it turned out. They were booked on the same flight to Dutch Harbor, both to work for the FCA. When they got to the hotel, they found out they'd be sharing a room.

Julio was impressed by the Grand Aleutian. The lobby was striking, with a towering stone fireplace. From his room, he could see big ships sailing in and out of the bay. There were two restaurants and a bar with its own menu. The men were allowed to order anything they wanted and charge it to the company. Julio was amazed at how well they were being treated. He ordered steak and lobster. The only things that were off-limits were booze and porn. There'd been some problems in the past, they were told. If you want to drink, do it with your own money.

It was cold and snowing. Neither of them had winter boots, but Julio and Rosel left the hotel and wandered down to the waterfront. There were some boats tied up, including the *Cornelia Marie* from the *Deadliest Catch.* They walked by a handful of storefronts. There was a barber shop, a liquor store, a tanning salon, and a bar—the UniSea. They continued a long way along a road that traced the bay. More than once, Julio fell; it was hard to balance on the crusted ice built up along the road's shoulder. But it felt good just to breathe in the cold, crisp air. It was a nice change from the overbearing heat in southern California.

Julio had been working hard for years but couldn't seem to save much money. He didn't own a home. He had no wife, no kids, no obligations. He was ready for a big change—an adventure. For three years, his cousin Marco had been flying up to Dutch Harbor for months at a time, coming home flush with cash. He'd said he might be able to get Julio a job, and maybe their cousin Byron, too.

Byron and Julio grew up together in Guatemala. Julio was three years old when his mother left for the United States. She got a job in a restaurant in Queens, New York. He was raised mostly by his grandmother. She was often caring for ten kids and struggling to make enough food to keep them all fed. Julio slept in the same bed with his younger cousin Byron, and the two boys grew to be like brothers. Byron looked up to Julio. When Julio became an altar boy, Byron wanted to be one as well. Byron was confirmed in the Catholic church shortly after Julio. When Julio was sixteen, he left home. Alone, he traveled the more than two thousand miles to the United States border. He took buses and sneaked onto trains. When he reached Tijuana, he called his mother, who had moved from New York to Los Angeles. She would pay for a coyote to bring Julio across.

He was the youngest in a group of sixteen. They crossed the mountains at night, hiking for hours by moonlight. There was one guy who was just a couple years older and had been back and forth before. Julio asked him about California: What was the food like? Had he been to Disneyland? He'd been there, he told Julio. It was great, even better than you could imagine. There was a van waiting for them on the other side. When they climbed out the next day, they were near downtown Los Angeles.

Julio enrolled in high school in Long Beach, but there was a lot of gang activity at the school and a lot of killings. He dropped out. He worked in the kitchen at a hospital, and then at a Ralph's supermarket. In 1995 he got a job at a mom-and-pop marine supply company that serviced boats at Marina del Rey, an upscale boat harbor just a couple miles north of Los Angeles International Airport. He had originally been hired for a construction job: The store was moving spaces and they needed some demolition work done. Julio's work ethic impressed the store's owners. They'd once hired their marine electricians straight out of the local high school, but that school had eliminated most of its shop classes. Julio became an apprentice. He learned marine plumbing and electrical work. There were a lot of wealthy people around, couples with fifty- or sixty-foot yachts who wanted custom work done. Julio worked on Florence Henderson's boat. He worked on John Travolta's boat. More and more of the shop's clients were Mexicans, and it was helpful for the owner to have a Spanish-speaking employee. After a few years, a lot of people came into the store and asked for Julio. For eleven years he'd lived and worked in Venice Beach. Byron was nearby. He had come north in 1987, three years after Julio. He was married, with two little girls. The cousins spent holidays together. Byron had learned to cook from their grandmother. He'd put on an apron and impress the family with his ceviche and carne asada. Sometimes, he and Julio talked about Alaska.

Julio had been thinking about it for a while. He'd done some online research and filled out a few applications for fishing jobs. Then, in the fall of 2007, his cousin Marco called. The Fishing Company of Alaska was hiring; Julio should contact their offices in Seattle, he said. The person who answered the phone at FCA wasn't interested in Julio's marine electrical experience. They'd be hiring new factory workers soon, though. Julio could come in for an interview and orientation. Then they would call him when they needed him.

Julio and Byron traveled to Seattle for the interview, then moved in with an uncle south of the city. Julio was following Marco, and Byron was following Julio. That was the way the family saw it, anyway. At Christmastime, with still no word from the company, Byron flew back to California. Julio stayed in Washington, waiting for the call. It came at the end of February. He was given a confirmation number for a flight the next day from Seattle to Anchorage and on to Dutch Harbor.

IT WAS 3:00 IN THE MORNING when the phone rang in Julio's room at the Grand Aleutian. They should pack their bags and come downstairs, the woman from the FCA said. Their ride would be there soon. Julio was the first one to the lobby. Soon he was piling into a van with half a dozen other new workers. It was just a ten-minute drive—across the Bridge to the Other Side, and right on Captains Bay Road. Then, the van stopped. The boat was tied up on the pier. It was much bigger than Julio had imagined. This is going to be fun, he thought, as he lugged his duffel out of the van and walked down the dock toward the waiting ship.

At 184 feet, the *Alaska Ranger* was almost twice as big as many of the vessels featured on *Deadliest Catch.* The hull was black, the wheelhouse a white, rectangular compartment perched on the front third of the ship and surrounded by a narrow upper deck with white metal rails. Julio saw his cousin Marco. They hugged, and Marco carried Julio's bag as they climbed the metal gangway onto the ship's deck and then down a flight of steep stairs and through the galley to the room they would share with six other men. There were four bunk beds, most of them strewn with crumpled sleeping bags, pillows, and clothing. Julio threw his bag on an empty mattress.

The *Ranger* wasn't the largest boat in the FCA's fleet. The company had the 200-foot *Alaska Warrior* and the 230-foot *Alaska Juris*. They owned seven boats in all. Two of the ships were long-liners, vessels that release massive fishing lines strewn with thousands of hooks into the ocean. The lines are left to soak and are pulled up hours later, the fish gaffed off one at a time as the lines are reeled back in. The other five FCA ships, including the *Ranger,* were bottom trawlers that target groundfish schooling near the ocean floor. The trawlers roll a huge, weighted net off the stern of the boat. The net is funnel-shaped, with a narrow rear pouch known as the "cod-end." The mouth of the net is held open by two refrigerator-size metal doors that are pushed apart by the force of the water as the net is dragged behind the ship, often for a dozen miles at a time.

Rollers on the underside of the net help prevent it from becoming caught up on rocks, coral, and other snags on the ocean floor. In a several-hour-long drag, 150 tons of fish can be scooped up in the net, which is hauled back on board with massive winches. The full cod-end looks like a giant sausage dragged up on the stern. The net is zipped open and the fish spill out on deck. They are shoveled through a hatch into an eighty-ton holding tank. Then they're sorted. Prohibited species like crab, salmon, and halibut, and bycatch like jellyfish, starfish, and other invertebrates, are thrown overboard. The "keepable" catch is headed, gutted, and stacked in gigantic freezers in the bow of the boat.

Marco brought Julio down to the factory, one deck below their bunk room. Julio glanced over the stainless steel tables and the silent saws. Marco showed him the enormous freezers. It seemed like they took up half the ship. The *Ranger* was a head-and-gut boat, the factory an assembly line for turning a freshly caught fish into a store-ready slab of flesh. A boat like the *Ranger* could decapitate and disembowel tens of thousands of pounds of fish each trip.

The vessel was one of about sixty such ships sailing out of Dutch Harbor. Most of these so-called H&G boats range from 100 to 225 feet, smaller than the Bering Sea's massive factory processors (the largest of which is more than 400 feet long and has a 100-person crew) but much bigger than most of the family-run boats that dominate TV images of Alaskan fishing life. Many of the smaller catcher boats that dock at Dutch and at the smaller Alaskan fishing ports of Kodiak, Seward, Sitka, Homer, Cordova, and Ketchikan, are just 30 or 40 feet long, with crews of two or four people. They'll fish for a day or two and then return either to port or to a floating processor to unload their catch while it's still fresh. The H&G boats, on the other hand, can hold enough fish to travel far from shore for weeks at a time. Most have multiple freezer holds built right into the hull of the ship. The *Ranger* had four freezers. On an exceptionally good trip, it might take just a few days to fill them. More commonly, the ship might be out for two, three, even six weeks at a time.

Most of the jobs on the head-and-gut boats weren't really about fishing. Jeremy Freitag found that out when he was hired for his first Alaska job, three years earlier. They were factory jobs—assembly-line work in the cold, wet belly of a boat lurching on the Bering Sea. It was a whole lot more like standing on the line in a meat-processing plant than a day of deep-sea fishing. Still, it wasn't a bad gig. Jeremy was twenty-two. He had seen the nickel ad about fishing in Alaska when he was nineteen, soon after graduating from high school in the small eastern Oregon town of Lebanon. He drove six hours up to Seattle to fill out an application. It was the first time in his life he'd left his home state. He got the job and landed in Dutch Harbor in the summer of 2005. The FCA was his third company. He had learned that once you've had a job on a boat, it's not too hard to find another one. He had spent the late summer and early fall of 2007 on the *Ranger.* The work was hard and the hours were long. But it was fast money. The starting pay was $50 a day, plus a bonus of four or five cents a case per processor. On a ten-day trip, the *Ranger* might get back to port with thirty thousand cases of fish. It could add up to a $1,500 bonus for an entry-level factory worker like Jeremy.

Only a couple of the guys on the boat were full-time deckhands. Everyone else worked in the factory: twelve-hour shifts, with six hours off in between. The factory crew was broken into three groups, with two of the three on at all times. In a couple weeks at sea, the men almost never got more than five hours of sleep in a row.

Jeremy had pretty much always been tired, but he didn't mind the work. The money was a lot better than at the mill job he had right out of high school. He figured that in just a few years he could save enough money to buy a house. He guessed that he'd need about $100,000—enough to get him something decent in the farm country of central Oregon. Nothing fancy, just a middle-class home. He didn't spend much money in Alaska—he slept and ate on the boat. And when he came home, he could crash at his parents' house. He had a good job and a goal.

In January 2008 Jeremy had planned to go back to Dutch Harbor to a spot on the *Seafreeze,* a three-hundred-foot processing ship he'd worked on back in 2006. He liked the larger boat better. The pay was a little more, the hours a little shorter. But when the time came, the *Seafreeze* didn't need him. He called the FCA. They had openings. In early January, he flew back up to Dutch, back to the *Ranger.* Jeremy had been thinking he'd work in the factory again, but the *Ranger*'s steward still hadn't shown up the day the boat was set to leave for the season's first fishing trip. The *Ranger*'s cook, Eric Haynes, asked Jeremy if he wanted the job. It paid $110 a day, a huge raise from his salary as a processor. Best of all, unlike a processor's twelve-hours-on, six-hours-off schedule, the steward worked days: 7:00 A.M. to 7:00 P.M. By FCA standards, twelve hours a day was the short shift.

Jeremy liked the work. He helped out in the galley, prepping meals and washing dishes for Eric. The ship's cook had been working on the boat for years, and he was good company. He'd been on the *Ranger* longer than the captain, longer than the mate, and longer than any of the engineers. Though he'd had offers from other companies, Eric liked the *Ranger.* He made eight or ten grand a month—only the ship's officers made more—and shared a room with just one other guy, assistant cook Mark Hagerman, with whom he traded off shifts in the kitchen. Eric kept a punching bag in his bunk room, and would work out in his off time. If the weather was good enough, he'd run wind sprints or jump rope up on deck. Back home in Las Vegas, Eric competed in amateur kickboxing competitions. He worked as a cook there, too (he'd once had a stint at the rotating rooftop restaurant in the Stratosphere). But after a while, he always wanted to return to Alaska.

The *Ranger*'s galley was about as basic as you could get, but Eric put effort into the cooking. Each season, he studied different cookbooks to perfect a new set of recipes. He knew how hard the guys were working; he'd started out in the factory himself. His food was their one pleasure—their reward. It made him happy to see them eat, and his effort didn't go unnoticed. The *Ranger* was widely regarded as having the best food of any boat in the FCA fleet. Most guys would say they ate a whole lot better at sea than they did at home. There were meals every six hours to coincide with the changing shifts, and the food was designed to please a crew made up of people from all over the world: tacos, stir-fries, sushi, steaks.

There was usually a pile of dirty dishes for Jeremy to deal with in the morning. After he took care of stuff in the kitchen, he would scrub the toilets, mop the floors, or maybe clean up the shower area a little. A couple of the showers had to be turned on and off with a pair of pliers or a crescent wrench. It was a pain how stuff like that never got fixed, Jeremy thought. The ship had two washing machines and two dryers, but one of each was broken, too. Keeping up with the laundry—mostly clothing stinking of sweat and fish guts—was a twenty-four-hour job. The crew joked that you didn't want to piss off the steward, or he'd wash your things without detergent. Jeremy was in charge of collecting and disposing of the ship's trash—pretty much everything was burned and dumped into the ocean. He would go around to the crew's rooms and collect their garbage. Sometimes he noticed beer cans or liquor bottles in the trash. Like everyone else, Jeremy had signed the company's no-tolerance drug and alcohol policy. "No illegal drugs, controlled substances, alcohol, paraphernalia, or firearms will be allowed aboard an FCA vessel at any time," his employee handbook read. "THIS MEANS NOT ONE SHRED, GRAIN, PILL, OR TRACE."

He knew booze wasn't allowed on board. But who was he going to tell? The bottles were in the rooms of some of the most senior crew. Jeremy had to go down to the engine room to get fuel for the garbage. On his way, he'd pass by the rudder room, where big, white absorbent pads the crew called diapers were laid out on the floor to sop up seawater and oil. The engineers would change the pads regularly, and Jeremy would have to haul big black trash bags stuffed with the sopping, oily diapers up to deck to be burned with the rest of the garbage. He'd go up to the wheelhouse to do odd chores for the captain, Steve Slotvig.

Slotvig was a competent captain and kept the boat running relatively efficiently, but there wasn't a lot of chitchat with the man. Small things could set him off, and his spirit was anything but generous. On Jeremy's first trip on the *Ranger,* one of the other processors got a nasty gash and needed stitches—a job that would normally fall to the captain of the boat. Slotvig was in no rush to deal with it. He wanted to finish his breakfast and coffee first. He told the man he could take one shift off—six hours. After that, he'd better be back on the factory line, or he was out of a job. Jeremy quickly figured out that he should go out of his way not to get on Slotvig's bad side. He brought the captain his meals and coffee—whatever he wanted. He washed the windows of the wheelhouse, scrubbing off the hardened bird shit with Windex and an ice scraper. It was hard work in the winter, when the temperature outside was regularly below freezing. Bird shit was bad. Frozen bird shit was worse. But the job still beat the factory. Jeremy considered himself lucky.

This was the first year Jeremy had been in Alaska for the

start of the winter A season, which stretches from early January through late March, sometimes into April. It was much colder than he'd experienced before. The waves were bigger. There was lots of ice in the water. He'd noticed back in the summer that the *Ranger* rode rougher than his previous boats. One of the ship's engineers had told him it was because of the *Ranger*'s flat bottom. It was built as an oil-rig supply boat for the Gulf of Mexico back in the early 1970s—a "Mississippi mud boat," they called it—and converted to a fishing trawler decades later. Jeremy just knew that it felt different than the ships he'd been on before. It seemed like a sloppy ride.

DUTCH HARBOR'S STAR ROLE in the American fishing industry is relatively new. Until a few decades ago, Alaskan waters were teeming with foreign vessels. Fishermen from Russia, Poland, Korea, and Japan all had discovered the richness of the North Pacific's continental shelf. American fishermen, on the other hand, were nowhere to be seen. Then, in 1976, Congress pushed the borders of the United States two hundred miles out into the open ocean. The legislation was known as the Fishery Conservation and Management Act, and was later renamed the Magnuson-Stevens Act, after the senators who championed the cause (Republican Ted Stevens, from Alaska, and Warren Magnuson, Democrat from Washington State). The new "territory" was called the Exclusive Economic Zone (the EEZ), and it limited fishing in the area to American interests. Now, only American-flagged ships with American captains can fish within the EEZ.

But the foreign ships had something the Americans did not have. To fish for weeks or months far from their own shores, foreign vessels had developed technology to process and freeze their catch right on board. American fishermen had little experience with that type of fishing or with the Asian fish market and its insatiable demand for odd fish and—to the American palate—even odder fish products. And so, in the years after the Magnuson-Stevens Act was passed, U.S. companies hired foreigners—specifically Japanese—to teach them how to fish the Bering and to process fish products with specific Asian markets in mind.

The FCA was one of the first American companies to start processing on board their ships, and from the start they sailed with senior-level Japanese crew. At the time, it wasn't an uncommon practice. What was unusual was the company's Japanese family history. FCA is owned by Karena Adler. Now in her late fifties, Adler founded the company in the mid-1980s, soon after divorcing from Masashi Yamada, a Japanese businessman twenty-nine years her senior. Yamada, who is now in his eighties, remains a powerful businessman in Japan. He's involved in real estate, manufacturing—and fishing. Among his many holdings is controlling ownership of Anyo, a fishing company that operates five of its own factory ships, and is the exclusive buyer of all of FCA's catch. Anyo calls the Fishing Company of Alaska its "partner" company. The FCA doesn't have its own Web site. Until recently, though, its boats and their whereabouts were detailed on Anyo's home page.

JEREMY NOTICED THE DIFFERENCE that the Japanese made right away. On his other boats, the captains relied on sonar fishfinders and their own expertise to identify the best fishing grounds. On the *Ranger,* it was up to the head Japanese crew member.

Everyone called him "the fish master." Jeremy didn't know the fish master's real name, but he knew he was an important person on the boat. He often saw him alone in the wheelhouse, driving the ship. On Jeremy's other boats, only the captain or first mate did that.

The fish master had his own crew of technicians: There was a Japanese boatswain (or deck boss), a Japanese factory manager, and a few Japanese engineers. It seemed to Jeremy that the Japanese were running the show, and he was told by guys who'd been working for the company much longer that it was the same way on all the FCA boats. Like the captain and first mate, fish master Satoshi Konno had a private stateroom. The other Japanese technicians shared a bunk room, and they all usually sat together at meals, apart from the rest of the crew. Jeremy would help Eric Haynes prepare a separate meal for them, often fish, and always rice. Konno would frequently come down to the galley to request a particular dish. It often seemed like he'd been drinking; one time Konno even showed up in the galley with his pants on backward. Jeremy had heard that the Japanese didn't work directly for the Fishing Company of Alaska. In fact, they got their checks from North Pacific Resources, a subsidiary of Anyo, the Japanese company that bought the fish. None of them spoke much English. They spoke Japanese to one another, and when the fish master was communicating with the captain, there was more yelling than talking.

The captain was a screamer, and Konno got on his nerves more than anyone. They argued constantly. Jeremy saw it in the summer, during his first FCA trip, but recently the arguments had taken a more serious turn. The Arctic ice edge that normally pushes into the Bering a few hundred miles north of Dutch Harbor had crept unusually far south this winter. The seasonal ice is often broken into pancakelike chunks that can be two or three feet thick and, in certain wind conditions, migrate a mile in an hour. As January turned to February, the crew watched the captain and fish master butt heads about how fast the ship should be operated in ice.

Ice is everywhere during the winter in the northern Bering Sea. It's not uncommon for Alaska's northernmost harbors to become iced in during the winter months. The Arctic ice edge is moving all the time—a fishing ground that one week is beneath clear ocean may be iced over a few days later. Few fishing boats out of Dutch Harbor are icebreakers. The *Ranger* certainly wasn't. Still, most boats will occasionally find themselves fishing near ice, or trying to outrun the thickening pack. How deep they go and how hard they push is a judgment call on the part of the captain.

Anyone who'd been on the *Ranger* for a while had seen plenty of ice. Sometimes there would be barely any open water visible at all, just big, flat slabs of sea ice with cracks dividing them. The ship would usually just nose into it at a knot or two. The goal was to keep the speed slow, not get stuck, and avoid backing down, which could damage the ship's rudders or propellers.

This season seemed different, though. The ice was dense, and there was a lot of scraping and banging. It seemed to several of the crew like they were pushing through the ice faster than they had in the past, on this or on other boats. The *Ranger* had a walk-in freezer in the bow of the ship, on the same deck as the galley. Eric would send Jeremy up there all the time to grab stuff for the kitchen. From inside the freezer, it felt like the 184-foot steel-hulled boat was a pinball, banging fast and hard against one chunk of ice after another. They were definitely going slower than they would at full speed, Jeremy thought, but the pounding and vibrations were still startling.

The crew had been told that their numbers were lower than

on the other trawlers, that they were catching fewer fish and processing it more slowly. The fish master was angry with the totals and would take it out on the captain. Meanwhile, Slotvig would complain to anyone who would listen that Konno was going to get him fired.

Cook Eric Haynes was sick of it. Both Slotvig and Konno would come to him, complaining about the other. The captain would be whining about how the fish master had yelled at him or thrown something at him. Konno, meanwhile, would tell Eric that the captain was *baka*—stupid, or, his English translation: "small head." The fish master had been on the boat for a few years longer than Slotvig, and Eric had seen him interact with previous captains. Konno wasn't easy to get along with, period. The guy who had been captain before Slotvig left the company after a blowup with him. A hydraulic line broke and oil got all over the fish in the fish bin. The fish master wanted to pack it anyway. The captain at the time made the crew throw the contaminated fish overboard. He got his way, but he and Konno almost got into a fistfight over it. When they got back to port, that captain was gone. He left the ship and the company. Eric hadn't seen him since.

Konno was just so damn competitive about their numbers, Eric thought. It killed him when the other boats were doing better. And his personality and Slotvig's clashed. They were both short-tempered, and neither was the type to let anything go. To Eric, some of it was pretty immature. One time the previous year, the ship's steward had come to Eric to tell him about something the captain had just written on a dry erase board in the galley.

"Look, Steve just came down and wrote this big thing out there, that he's fired. That Konno fired him," the steward said.

Eric went out and erased it all. Then he went to talk to Slotvig. The man was in tears. The ship's cook ended up mediating the conflict between the two men, going back and forth between them to settle things down. It blew over, but the captain and the fish master hated each other to the point that Konno wouldn't even talk to Slotvig anymore. Eric had seen Konno get up and leave when the captain entered the galley. It was ugly, and it was playing out in front of the whole crew.

One night in late February, a handful of crew members were in the galley when they heard a commotion in the hallway one level up. Through the open stairway, the men could see the captain and the fish master standing chest to chest, huffing and puffing at each other. The men in the galley grew quiet. It was an argument about the ice. The captain had evidently been asleep, and had woken to find the fish master had increased the ship's speed—even though they were still inside the ice pack. Slotvig ran up to the wheelhouse and reduced the speed. The men in the galley had noticed the boat slow down. Moments before, it had been difficult to stand up without holding on to something. It felt like the ship was almost at full speed, the hull beating into the ice pack with a noise like a pounding drum. Now, the captain was screaming at the fish master, hollering that plowing through the ice that way could damage the boat. The captain was fuming.

"It's *my* fucking boat," several crew members heard him yell at Konno. "I'm the one that drives it!"

The fish master was cussing back at the captain in Japanese. The men didn't understand a word he was saying, but he was obviously enraged. They could see the two men shoving at each other in the narrow hallway.

A couple more crew members had stuck their heads out of their staterooms before the fish master spit right in the captain's face.

The next time the *Ranger* arrived back in Dutch Harbor was the day Julio Morales boarded the ship for the first time. Julio's first job had been to keep an eye on the vent on one of the fuel tanks as the ship was pumped full of diesel. He was posted on the deck on the port side when he heard screaming from inside the wheelhouse.

"Get the fuck out of here, motherfucker!" Julio heard someone yell. He had no idea who it was, or what was going on, but the words were clear and angry: "Get the hell out, you motherfucker. Go home!" Whoever was yelling had a strong accent. Julio could hear slamming from inside the ship. About five minutes later, he saw Captain Steve Slotvig emerge, carrying a garbage bag stuffed with clothes. The older man looked pissed as he stormed down the pier. Julio heard the crew talking. They were saying that the fish master had gotten the captain fired.

By the time the *Ranger* left Dutch the next day, a temporary solution had been found: Eric Peter Jacobsen, a twenty-plus-year FCA employee who'd been serving as first mate under Slotvig, would fill in as captain. Most of the crew was relieved. Unlike Slotvig, Jacobsen, whom the crew called "Captain Pete," was well liked and mild-mannered. His first mate would be David Silveira, a fifty-year-old former tuna man from San Diego who was the captain of one of the FCA's long-liners, the *Pioneer.* Silveira wasn't happy about the assignment, but he was willing to step in temporarily. Pete was a friend. Both men were used to sailing on long-liners, and Silveira didn't want to leave Pete to deal with the trawler and its crew on his own. He agreed to go—but just for a few trips, just until they found someone else.

ON MARCH 5, THE *ALASKA RANGER* left Dutch Harbor with Pete Jacobsen as her new captain, and David Silveira as mate. It took less than a day to reach the fishing grounds. They were targeting yellowfin sole, but a half dozen species were piled up around Julio's new knee-high rubber boots: pollack, rock sole, cod, halibut. Not that Julio could tell what was what. He quickly realized there wouldn't be any formal training in this job. No apprenticing like his work at the marinas in Southern California. He had to watch what everyone else was doing and figure it out, quick.

Julio's job was to wade into the fish bin, which had been loaded up through a big hatch on deck, and push the fish out onto the conveyer belt on the ship's port side. The fish ran over a flow scale that recorded the total weight of the catch, and then were herded into a hopper where the bycatch and "prohibs" were sorted out. A stream of salt water kept the fish wet and slippery. Dry fish were too hard to move around. The water soaked the whole factory; sometimes there would be several inches of standing water on the factory floor.

From the hopper, the fish were fed out onto two conveyer belts, and a couple guys stood on the belts and kicked the fish into the right direction so they'd be lined up head first when they reached the circular saws at the ends. Those guys were called the "kickers." The "headers" manned the saws that decapitated each fish. The fish needed to be lined up neatly to keep as much of the flesh as possible. After the heads were off, a mini-vacuum attached to the saw sucked the guts from the body cavity. The detritus was fed out a discard chute back into the ocean; the "shit chute," the guys called it. Then the fish were sorted by species and size, packed into metal pans, and stacked into a plate freezer that squeezed the pans together and compressed the fish into compact blocks.

It was exhausting work. It was cold in the factory, and loud, with all the noise from the conveyor belt gears added to the constant grumble of the *Ranger*'s massive 7,000-horsepower engines. Julio didn't know too many names, and no one had learned his, either. People just yelled "Hey, hey!" to get his attention. He was learning some of the common language though, like the "hubba, hubba" urged by the Japanese techs. It meant hurry up.

There were video cameras in the factory. That was so the fish master could keep an eye on production from up in the wheelhouse, Julio was told. Sometimes the Japanese boss came down and walked the processing deck. He'd grab smaller fish off the line and throw them out. It seemed to Julio like a lot of the men were nervous in the fish master's presence. He could be ruthless in chewing out an incompetent factory worker and no one wanted to piss him off, or look lazy in front of him. When the fish master got close, Julio could smell alcohol on his breath.

From the end of Julio's first six-hour shift his back hurt and his arms were sore. He didn't complain. Don't think about it, he told himself. Just keep working. When the men went up to the galley for lunch, they stripped off the rain gear they wore in the factory, which was covered with little bits of fish, scales, and slime. They hung it on hooks, wiped their hands on the sweatshirts they wore underneath, and went to eat.

After thirteen days, the boat was full with thirty-two thousand cases of fish. When they got back to Dutch Harbor, Julio was happy to see his younger cousin Byron Carrillo at the dock. Byron had gotten a call just a few days after Julio. He was originally assigned to a different FCA boat, the *Juris,* but when Byron got to Dutch Harbor he asked if he could be put on the same ship with his cousins.

On Wednesday, March 19, the *Ranger* headed back to the yellowfin grounds. They were up near the Pribilof Islands, a couple hundred miles north of Dutch Harbor, when the first haul came up. Byron had been assigned to a different work squad than

Julio and Marco. When their shifts overlapped, Julio could see him from across the processing space. His cousin had thick, shoulder-length black hair that was flapping all over his face. He had brought way too much stuff to Alaska: three bags in all, with a bunch of nice clothes and shoes he'd never need in Dutch Harbor. "This isn't a cruise," Julio told Byron when he saw the bags. Even with all that gear, Byron hadn't brought a hat. Julio had an extra knit beanie. He gave it to his cousin to hold back his hair.

After only a couple of shifts, Byron was exhausted. It seemed lucky when it turned out his first trip was an unusually short one. The ice had shifted when they had been back in town and now solid pack covered the intended fishing ground. They dropped the net, but the water was too deep and they weren't hauling up the fish they wanted. On Friday, March 21, the FCA decided to cut its losses, and ordered all the trawlers to return to Dutch.

The weather was terrible back in town. Twenty-five-mile-an-hour winds chapped the men's exposed faces as they secured the ship to the pier. They would be going right back out, but there was still time to make a few phone calls. It was after midnight as Julio and Byron shuffled down the icy pier to a bank of pay phones. There was an arctic fox digging into a nearby trash can, and a few guys lined up to make calls. They both talked to Julio's brother, who was about to be deployed to Iraq. Then Julio waited in the snow while Byron called his wife and young daughters in Los Angeles.

BACK ON THE SHIP, THERE WAS still plenty of work to be done. Pallets of groceries were delivered to the dock, and steward Jeremy Freitag helped load the food onto the boat and unpack it into the *Ranger*'s walk-in freezer. Julio and Byron were part of a group of crew members tasked with loading on a month's worth of "fiber," the bags used to hold the frozen fish. The crew formed a chain from the dock to a storage space next to the wheelhouse, handing bundles of the sacks along like men on a fire brigade. After just a few loads, Byron bent over and asked Julio to pile the bags on his back. His arms were too tired, he couldn't hold them up anymore. In Los Angeles, Byron had been working as a cashier at a gas station, and at a pizza restaurant. He wasn't used to hard physical labor. He was a little overweight. And he really didn't like the cold. After the first day in the factory, Julio had asked his cousin if he thought he would come back for the summer B season. "Hell no, I'm not coming back," Byron said. The work was miserable.

While most of the crew were busy getting the ship loaded up for the next trip, the deckhands concentrated on changing the fishing gear. They would need the bigger net. Everyone had heard that all of the FCA's trawlers were heading out to fish for Atka mackerel, a foot-long, bumblebee-colored fish that schools near the edge of the continental shelf and is popular in Asian cuisine (most Americans have never tasted the fish).

The targeted fishing ground was four hundred miles west, halfway to Russia, in an area known as Petrel Bank. On Saturday morning, March 22, the ship's engineers fueled the *Ranger* with almost 150,000 gallons of diesel. It was noon by the time the ship pulled back out into Captains Bay and began the long steam out the Aleutian Island Chain.

Chapter Two

"Go to the Suits!"

Matt Duben could see the air station's runway from his living room window. It was Saturday afternoon, March 22, and the snowplow was out there, slowly clearing the strip.

Duben flew the Coast Guard's fixed-wing airplane, the Hercules C-130, or "Herc." He was forty-five and this was his second Kodiak tour. He and his wife had been stationed on the island in the mid-1990s. They loved it so much they had come back on vacation almost every year since. The town was custom-made for their lifestyle: hiking, fishing, hunting. And the flying was spectacular—arguably the most challenging flying any pilot could ever do. The environment was harsh and there were endless logistics associated with the long distances and violent weather. All the cold, snow, and ice made mechanical problems common. But it was so beautiful. On prettier days it couldn't get any prettier, and on nasty days you couldn't imagine anything nastier. It was his kind of place.

Duben and his wife had five-year-old triplets, a girl and two boys. He'd taught them to fish on a local river. They lived right on base. When he was on duty—typically one or two twenty-four-hour shifts a week—he could wait at home for the call. The goal was to be airborne within half an hour of when a search and rescue (SAR) case came in. From his house, Duben could jump in his truck and be at the hangar in less than five minutes.

The weather report at that morning's air station briefing wasn't alarming but Duben knew well that the weather in Kodiak often doesn't align with the forecast. He'd learned to look out the window and gauge the conditions on an hour-by-hour basis. By midafternoon, it was snowing pretty heavily, with serious wind gusts.

The runway at Kodiak is short, and the wind direction makes even a routine takeoff difficult. In the winter, Kodiak C-130 crews regularly predeploy to Elmendorf Air Force base in Anchorage, where they will be better positioned to respond to a search and rescue call. The weather tends to be clearer there, and the runway is far easier to take off from and to approach than the stubby strip in Kodiak.

If we're gonna go, we better get out of here now, Duben thought as he watched the plow chug up the asphalt. We should take advantage of the clear runway. Duben's copilot that day, thirty-one-year-old Tommy Wallin, was thinking the same thing. By late afternoon they were consulting with the day's operations officer. Everyone agreed: Duben, Wallin, and their five-man crew would fly the Herc to Anchorage.

First, though, they had to defuel the plane. The C-130's standard load of fuel is 45,000 pounds (6,600 gallons)—enough for nine hours of flying, or maybe more if the pilots take extreme measures to preserve gas. But the heavier the plane, the longer the runway needed for a safe takeoff. A wet, slippery airstrip makes things worse, which was another reason to head to Elmendorf. If they got a SAR case, the pilots would try to take off in almost any conditions. But from Kodiak in bad weather, it would be impossible to take off with enough fuel to reach an emergency far out in the Bering Sea. It was better to preemptively head to Anchorage.

It was dusk by the time the plane lifted off above the snow-shrouded island. The flight east took just over an hour. As usual, the weather in Anchorage was far better, and after prepping the plane and refueling, the crew loaded into a van to a hotel on base. Elmendorf didn't have much room for the Coast Guard. The base didn't have an extra hangar for them, so the crew left the plane on the taxiway. There were beds, though. Thirty bucks a night in Air Force billeting. Before midnight, the entire seven-man crew was asleep.

EMPTY FOOD TRAYS WERE LINED UP in the *Alaska Ranger*'s galley, a few holding cold pizza slices hardened with congealed cheese. Every Saturday was pizza night. That was the one constant in Eric Haynes' weekly menu. It was 2:00 in the morning, but four or five guys were still clustered around one of the galley's tables—among them boatswain Chris Cossich and factory manager Evan Holmes. They were watching an old boxing video on the galley's TV, killing time while there was plenty of time to kill.

The ship had been under way for almost fourteen hours, steaming west toward the fishing grounds. There wouldn't be any factory work until the first haul came up. In the meantime, the crew cherished the downtime. They watched movies and TV shows:

DVDs of *South Park, Family Guy,* and *King of the Hill*. Between meals, some played poker, a few read. Mostly they just slept.

The *Ranger* was already more than one hundred miles from Dutch Harbor. Every FCA trawler was headed to the Atka grounds. The *Ranger* had left Dutch hours ahead of the others, though. Fish master Satoshi Konno was eager to get going. As the ship steamed out of Captains Bay just after noontime on Saturday, March 22, the other FCA trawlers—*Alaska Juris, Alaska Spirit, Alaska Victory,* and *Alaska Warrior*—were still tied up at the pier.

Evan and Chris were two of five men on the *Ranger*'s e-squad, or emergency squad. A couple years before, they had each gone to a week of basic safety training in Seattle. The company had suggested it and paid for it. The class covered everything: firefighting, man overboard, abandon ship. They jumped off a pier into Puget Sound and practiced climbing in and out of life rafts. They learned CPR. Evan thought it was a really good class.

He was twenty-five years old and from Sonora, California. He'd been working on the *Ranger* for just two years, but had moved up fast. As a new processor, he'd tried hard to impress people, and his effort had paid off. Just a few days before, Evan had been promoted to factory manager, one of the more important jobs on the ship. It didn't hurt that there was such a high turnover. The previous summer B season there'd been at least a dozen brand-new guys all starting at once on a forty-five-man crew. This year they'd had about an equal number, though spread out over the season. Four or five new guys had started in just the past week. If they were hard workers and stuck around, they'd move up quickly, too.

Evan was pretty sure that the fish master had played a role in his promotion to factory manager. Konno wasn't friendly with many Americans, but he seemed to like Evan. He was one of the only nonofficers whom Konno knew by name. "Holmes!" he'd yell. Some guys had been on the boat for two or three times as long as Evan and still weren't even a shift leader. The fish master liked to give people like that a hard time. One time Evan had been sitting with another crew member when Konno pointed at the guy.

"How long on boat? How long fishing?" the Japanese man asked.

"Eight years," he answered.

Konno turned to Evan: "How long?"

"Two years."

"Ha, ha! Nice," the fish master mocked.

"Dipshit," the other guy said.

Konno wasn't an easy person to like, but Evan figured getting along with him was basically a matter of doing his job well. The man was all about getting the fish, and he was no hypocrite. In freezing weather, when everyone else was bundled up and moving slowly, the fish master would be sprinting around handling the net with bare hands. He was faster than anyone else. Konno was a bust-ass, no-bullshit worker and he expected the same of the people working under him. That was something to respect.

EVAN WAS RECLINING ON A BENCH across from the TV when the galley's A-phone rang. The A-phone system allowed the crew to communicate between the galley, the bridge, the engine room, and many of the officers' staterooms. Evan got up.

"Hello?" he said into the plastic receiver.

Nothing.

He hung up, but as soon as he did, the phone rang again.

"Hello?" He waited. "Hello?"

Still, no one.

"What the hell's going on?" Evan said aloud, but the other guys only shrugged. He left the receiver off the hook and started up toward the wheelhouse.

The ship's top officers worked twelve-hour shifts, with the captain and chief engineer on days, and the first mate and assistant engineer on duty at night. First Mate David Silveira was in the wheelhouse.

Silveira saw Evan approach the door with boatswain Chris Cossich right behind him.

"We're taking on water in the ramp room," Silveira said. "Go down there and do your jobs."

Evan and Chris looked at each other, then ran two flights back down to the galley.

"Hey, we're flooding!" Evan yelled at the rest of the guys in the dining area. They were still watching the old boxing tape. Evan and Chris kept moving deeper into the ship, one more deck down to the *Ranger*'s ramp room, at the stern of the boat. The room was basically the ship's shop, where the crew kept tools and supplies for repairs. It was called the ramp room because of the platform at the rear, where the *Ranger*'s massive trawl nets were pulled up.

Evan opened the door.

Holy shit, there was a foot of standing water in there.

He felt the breath sucked out of him.

Evan knew there should absolutely never be water in the ramp room. The space was on the second level of the ship, a full floor above the engine room, and rudder room. If you left a hatch open maybe a little bit of water could have splashed in from the factory, which was in the middle of the ship on the same level. But Evan had never seen it happen. Besides, the factory was shut down right now. Everything should have been sealed up and all the water lines turned off.

Evan could hear another A-phone ringing. Maybe several phones were ringing. He didn't hear an alarm going off, though. What the hell, Evan thought. It must have taken a while for the water to get this high. Where was it coming from? Evan had no idea, but as he stared at the far wall, he thought he could see the waterline slowly rising.

The pumps! There were two dewatering pumps on the far side of the shop. They'd have to wade through the water to get at them. Evan was wearing tennis shoes. Chris had on flip-flops. They looked at each other. Okay, this is serious, Evan thought. But how serious? He turned to Chris. "We gotta go," he said. There was no time to go back upstairs to change into their boots.

As soon as Evan stepped into the standing water his feet and calves began to ache and cramp up. The freezing water felt like a million pinpricks over his skin. His heart raced as they waded through the salt water, pulled a pump from a far wall, and struggled with it back across the ramp room.

We gotta get this thing going, Evan thought as they placed the pump down by the ramp room door. It took just a minute for the two e-squad members to get the ship's fire hose hooked up to the pump. Then Evan headed toward the stairway up to the deck, the long end of the hose looped around his arms.

Just as he was headed up the narrow metal staircase, Evan ran into Chief Engineer Dan Cook, who was coming down. Evan didn't know Cook that well. The chief was a big man, six foot two and close to 280 pounds, with a shiny bald head and a Santa-like white beard. The fifty-eight-year-old engineer had only been on the *Ranger* for a couple of months, but Evan could tell he was a jokester from the few times they had talked. Now the older man was deadly serious.

He stopped Evan on the stairs. "No, go to the suits," he said.

"What?" Evan said. He had the fire hose in his hand and was

prepared to bring it up to the deck where he'd run the end of the hose overboard.

The chief was breathing heavily and sweat was running down the sides of his bald head. "Get everybody up there, too, and get your suits on. We might have to abandon ship."

JULIO MORALES JOLTED AWAKE. An alarm was going off, a loud, ringing sound like an old-fashioned telephone. He looked around the dark, eight-man bunk room. Most of the racks had men in them; only a couple of people were moving.

It must be a training drill, Julio thought. Then, he heard someone yell from lower in the boat: "The rudder room is flooding!"

Julio climbed down from the top bunk. Out in the narrow hallway, he saw one of his bosses, Evan Holmes, running toward him.

"Get ready to abandon ship!" Evan yelled.

"Are you serious?" Julio said. Evan was opening stateroom doors, yelling people's names and shaking men awake.

"We got to get out," he screamed. "The rudder room is full of water!"

ERIC HAYNES WAS ALSO ASLEEP in his bunk when he was startled awake by his cabinmate, Assistant Cook Mark Hagerman, knocking at the door. "Eric, I don't know how bad it is, but we're flooding," Mark said. "We're taking on water."

Eric pulled on his sweats, grabbed his phone and wallet, and started for the wheelhouse. Captain Pete Jacobsen was up there, along with First Mate David Silveira, and Chief Dan Cook.

"What's going on, Pete?" Eric asked.

"We're taking on water in the rudder room," the captain said. He looked at Eric. "We're gonna go down. We're gonna sink."

What? Eric thought. What could have happened? The weather wasn't that bad—certainly not bad enough for water to have flooded into the ship from above. They hadn't hit anything. He hadn't heard anything.

Captain Pete was serious, though, and there was no time to ask questions. Eric rushed back belowdecks. He had seen the *Ranger*'s crew ignore the general alarm before. In his time on the ship, they'd had more than one Freon leak. The gas is deadly, but some guys had stayed asleep in their bunks right through the emergency alarm. Eric started knocking on doors, waking guys up.

"It's a real emergency!" he yelled. "Go up to the wheelhouse!"

The phones were ringing nonstop and the general alarm was broadcasting a single, loud ring throughout the ship. Plenty of people were still slow to get moving, though. One Japanese crewman was in his bunk with the curtain closed when Eric burst into the room.

"Everybody up top!" Eric yelled. "Right now. Move. Move!" The man pulled back the curtain and looked at Eric with a sleepy smile. "Let's go! You gotta go," Eric urged.

But when he ran back through several minutes later, the Japanese man was still lying in his bunk.

"Get up," Eric yelled again.

He didn't tell the men what Captain Pete had said. He didn't want to start a panic.

JULIO FOLLOWED EVAN HOLMES down one level to the galley, though the laundry room, and out onto the *Ranger*'s main trawl deck.

"Go to the wheelhouse!" Evan was yelling. "Go to the suits!"

Julio climbed to the ship's upper deck, where the factory manager began pulling bags containing full-body, neoprene survival suits out of a plywood box on the side of the wheelhouse. He handed one to Julio. The bulky, bright red suits looked a little like children's footed pajamas. Despite their awkward appearance, the suits provide insulation and buoyancy and—when worn properly—will keep someone dry, which is essential to retaining enough body heat to survive in cold water. Each suit was folded up inside its own bag, which was color-coded to indicate the size. Inside the bags, the neoprene survival suits were stiff from the cold.

On his first day on the boat, Julio had been shown the suits, and practiced putting one on. You were supposed to do it in sixty seconds. The suit had a fitted hood, and a flap to cover your nose and mouth that was held shut with Velco. Julio had been told that the most important thing was not to let water get inside.

The deck of the ship was coated with ice and so Julio brought the suit into the crowded wheelhouse. He shook the survival suit out of its bag, unzipped it, laid it on the floor, and yanked off his rubber boots. He wiggled his legs into the suit, then pulled the torso up over his sweatpants and FCA sweatshirt. It was pretty easy, but as Julio yanked the neoprene suit up, he noticed a rip at the seam on the left arm. There was another small tear at the suit's left ankle. This isn't good, he thought. If we go down, I'm in trouble.

There were other guys waiting to get into the wheelhouse to put their suits on. Julio went outside. It was snowing and windy. Only his face was exposed, but Julio was still cold. He stared back toward the *Ranger*'s trawl deck. It looked lower than normal. He watched as a big wave crested over the stern. Jesus, that's bad, he thought. Everyone seemed calm, though. This ship has huge bilge pumps. They should be working now, Julio told himself. They'd just wait, he thought. And things would be okay.

WHEN HE GOT BACK UP ON DECK after waking up the crew, Eric Haynes saw Chief Engineer Dan Cook stopped at the top of the wheelhouse steps. Cook was bent over, with his hands on his knees. He was breathing hard.

"Are you all right?" Eric asked. Cook told him he was just tired from running up from the engine room. The chief had been in bad health, Eric knew. Early in the winter, he'd come down with pneumonia and was sent home to San Diego. When he came back a few weeks later, it was obvious he wasn't fully recovered. Cook had been hacking and coughing constantly; twice a day he slipped away to his stateroom to use a special breathing machine. He didn't seem good, but there wasn't much Eric could do for him at the moment.

Just about everyone was in their suits already. Eric grabbed the last one in the box on the starboard side and then headed back belowdecks to put it on. Most guys were getting into their suits in the wheelhouse, but Eric could see it was packed in there. The suit was tight; he needed help from another crewman to stretch it around his broad shoulders. A few minutes earlier, Eric had passed a brand-new guy who had showed up at the dock just a couple days before. He was standing still, with his suit hanging down around his thighs. It was obvious the guy had no idea what to do.

To comply with Coast Guard regulations, the ship's crew was required to run extensive drills at least once a month. In the meantime, new crewmen were supposed to be given a primer on the boat's survival gear, and how to respond to an emergency on the ship. The last few guys who'd arrived hadn't gotten that instruction. Come to think of it, Eric realized, the drills they'd been doing lately were much less intensive than what they'd done in the past. Most often, the men just held their suits during the drill. It was too much trouble to take the suits out of the bags and get them folded neatly inside again. That was the thinking at least. And Captain Steve Slotvig had been adamant that only the ship's officers would need to launch a life raft. Every man on the ship had been taught how to do it in the past. Not lately. Plus, the fish master was always so impatient to get back to the fishing grounds, that it seemed they never took the time for a full drill at the dock anymore. Eric had heard Pete Jacobsen complain about it more than once.

Eric had known the sixty-five-year-old captain for years. The *Ranger*'s cook had first signed up with the FCA almost fifteen years before. Pete had been there much longer. He was fully qualified as ship's captain, but Eric knew Pete usually preferred sailing as mate. He didn't like dealing with the fish masters. As second in command, he made the same money, but didn't have to deal with power struggles with the Japanese.

Pete had first come north in the 1970s, working for the tugboat company Foss. A decade later, he was hired on as one of the FCA's very first employees. Seafaring was in Captain Pete's family from way back. He wasn't really a big talker, but over the years a few crew members had heard some stories about his history back East.

Both of Pete's grandfathers had worked on whaling ships in Denmark before emigrating to the United States, his maternal grandfather as a sea captain and his paternal grandfather as a chief engineer. Pete's father, Hans Jacob Jacobsen, became a ship's engineer as well. Pete was the second youngest of six siblings, all of whom grew up with the family's sea stories in Weymouth, Massachusetts. Pete's parents split up when he was young. He left school, got a job working at a nearby shipyard, and married his childhood sweetheart, Marie Allen. By his mid-twenties, Pete was the father of two young children, Karen and Carl. He named his son after his older brother, who'd followed their father out West a number of years before. His dad sounded content in Washington State. There was plenty of work out there for a ship's engineer. Back in Massachusetts, Pete was working two or three jobs just to pay the bills. He felt like he couldn't get ahead. Two of his four brothers, Carl and Billy, had moved West not long after his dad, and both had found good work in Seattle's maritime industry.

Pete was going on a trip to visit his brothers, that's what he said anyway. His family dropped him off at Boston's Logan Airport. They stood at the gate, waving good-bye as he boarded the plane. A week passed, then two.

"When's Dad coming back?" Karen asked her mother.

"Soon," she told her.

It was 1973. Karen was nine years old. After a few months, she stopped asking. A year later, the divorce papers came in the mail. Karen watched her mother cry as she opened the envelope.

Pete sent money and sometimes letters. Karen would write in return, mailing drawings she'd made. Her dad sent back art supplies and, once, two little Eskimo dolls dressed in real fur that Karen loved to rub against her face. One had a baby on its back, hidden deep in the thick pelt. He'd gotten a job up in Alaska, he wrote to Karen. It was so beautiful there, he said. He called it "God's country."

Karen missed him. She'd study her memories, and soak in the moments of her father's attention. She wondered if the shipyards out West were anything like General Dynamics in Quincy, Massachusetts, where her father had worked as a sandblaster before moving away (the other men called him "Jake the Snake" for his ability to squeeze into tight spots). When she was a little girl the whole family went there one morning for a christening. Karen remembered the champagne bottle crashing against the bow of a ship her father had helped build. They sometimes went to the drive-in theater in Weymouth, where Karen rode a little train that circled the parking lot. Afterward they ordered clam fritters to eat in the car during the movie. She and her little brother curled up in the back, almost asleep by the time the cartoons had finished and the feature film got started. She'd remember how they used to go to a local pancake house on the weekends, or how her dad would pick up a box of doughnuts and the *Boston Globe* and she'd cut out paper dolls while he read the funnies across their kitchen table.

When Karen was a sophomore in high school she went to visit her father in Everett, Washington, a suburb north of Seattle. By that time he'd been married and divorced a second time and was working most of the year in Alaska. The rest of the time he lived with his brother, Billy, in a bachelors' apartment in Everett. Karen attended the local high school for a couple months. She felt like she and her father had a lot in common. They were both morning people, both crazy about animals. On that trip, she loved to get up early with him to walk a little dog he had at the time, a Blue Heeler named Andy. He was busy, taking nautical classes, studying for his mate's license. He had a new girlfriend. But he found the time to teach Karen how to drive.

Put your blinker on even when you're in a parking lot, Pete Jacobsen instructed his fifteen-year-old daughter, even if there's nobody else around. You want to build good habits, he said. "Your job as a driver is to make the ride comfortable for the passenger." Years later, Karen and her mother would repeat the words to each other when one lurched too quickly in or out of traffic. Pete had taught his ex-wife to drive, too. Pete's lesson was a happy memory they shared of him.

* * *

AFTER ERIC HAYNES HAD HIS OWN SUIT ON, his focus turned to the muster sheets. He went back up to the wheelhouse, where lists of the crew were sealed inside a plastic pouch that was taped to the wall. Years ago, Eric had been part of the *Ranger*'s e-squad. Not anymore, but he was still on the muster team. With his hands encased in neoprene, Eric couldn't easily rip open the plastic pouch. He struggled with it for a minute, then grabbed a pen, and stabbed into the plastic envelope. He pulled out the sheets, which divided the names of the forty-seven-person crew into three groups, one for each of the *Ranger*'s twenty-man life rafts.

The muster groups roughly corresponded to the *Ranger*'s three factory shifts, and the crew was already gathering in the proper groups out on deck. The names on the sheets, though, were two trips old. Since they'd been printed up, at least half a dozen guys had left, and an equal number of new crew had boarded the ship. Eric handed out the sheets, and each group mustered as best they could. Then most of them crowded back inside the wheelhouse.

Evan Holmes looked around. Okay, most of these guys have their suits on pretty good, he thought. When they'd drilled in the past, there'd usually be a couple people who wouldn't want to take their beanies off. You couldn't have a hat on and get the hood of the survival suit sealed properly around your face. Evan had to tell one of the Japanese techs to take his off. Overall, the situation seemed relatively calm. Even if this thing sinks, there's probably time for another boat to get here, Evan thought. The officers were saying the *Alaska Warrior* was just a few hours away.

"MAYDAY. MAYDAY. MAYDAY. This is the *Alaska Ranger*." It was 2:46 A.M. when First Mate David Silveira picked up the HF radio and called the Coast Guard. Evan heard him repeat the ship's coordinates and listened in as the Coasties answered back.

"*Alaska Ranger,* this is COMMSTA Kodiak," the voice responded over the 2182 frequency. It was watchstander David Seidl. He was collecting the information the Coast Guard would need to launch an effective rescue mission to a spot almost eight hundred miles away, across a huge expanse of black ocean.

"Roger, good copy on position. Understand you are flooding, taking on water from the stern. Request to know number of persons on board, over."

"Number of persons is, um, forty-seven persons on board, okay?" Silveira answered.

Everyone was talking. Many people were smoking. The wheelhouse was growing cloudy with cigarette smoke. Eric Haynes could tell that the captain was stressed. He needed the men out of his way. The *Ranger*'s long-time cook herded the processors out to the exposed deck, where they immediately had trouble keeping their balance on the iced-over metal platform.

Eric went back inside. Most people already had their suits on, but not the captain or the mate. Konno, the fish master, hadn't put his on, either. He was talking with his technicians. Just the day before, Konno had been showing Eric pictures of his home in Japan, where he had a wife and teenage children. He had pictures of his Japanese garden. There were rocks with holes in them that Konno had drilled by hand. It looked to Eric like a lot of work.

Eric noticed that Chief Engineer Dan Cook had made his way into the wheelhouse and was sitting in the captain's chair. Cook was talking with Captain Pete and Assistant Engineer Rodney Lundy.

"It can't be saved," Eric overheard the chief tell the captain. "We should abandon ship." It sounded like Dan was convinced there was no hope, while Rodney thought the *Ranger* might make it if the watertight doors held. David Silveira was still on the radio with the Coast Guard, answering questions. The Coasties had their position, but Eric knew the rescuers would be coming from far away.

OUTSIDE, THE DECK WAS SLICK WITH ICE. The bow was covered with snow. Almost forty men were gripping the rails, some talking, many managing to smoke cigarettes even with both hands covered in thick neoprene. With everything but their eyes covered up by the red survival suits, it was hard to tell one man from another.

Julio Morales gripped an ice-encrusted rail. It seemed to him that a half an hour had already gone by since the alarm went off and he'd woken up to all the yelling. It looked like almost all of the crew was already in their survival suits. Julio was just outside the wheelhouse, struggling to balance on the icy deck in his footed suit. There was a crowd inside, but he and most of the other factory workers had been ordered out. Now they were lined up against the metal railing, wondering what the hell was going to happen next.

Julio stared back toward the stern. It looked low. The waves were crashing up past the trawl winches. The back edge of the stern deck looked like it was almost to the water line. He looked again at the rips in his suit and scanned the boat for his cousins, Marco and Byron. He was particularly concerned about Byron, who had only been aboard the *Ranger* for four days. Julio remembered that his cousin had never learned to swim.

Julio wanted to find him, but it didn't seem like a good idea to move. The deck was so slippery. He'd just wait, he thought, for someone else to tell him what to do.

Chapter Three

Always Ready

It was just before 3:00 A.M. and Coast Guard pilots Steve Bonn and Shawn Tripp were sprawled out in the tiny pilots' lounge on St. Paul Island, locked in a late-night Xbox battle of Call of Duty 4. The men were on a barren, five-by-seven-mile speck of rock in the middle of the Bering Sea, the largest of five tiny islands collectively known as the Pribilofs. Outside, the wind whipped across the tundra, building a wall of snow against the room's single, narrow window.

During the winter crab fishing season, the Coast Guard predeploys helicopter rescue teams—two four-man crews comprised of a pilot, a copilot, a flight mechanic, and a rescue swimmer—to the island for two weeks at a stretch. Coast Guard command implemented the predeployment program more than a decade ago in response to a sky-high fatality rate among crab fishermen. The rescuers are on standby to respond to emergencies in the fleet, which plies the 32°F waters near the islands in search of opilio crab, a spindly, pale orange crustacean whose sweet meat is often marketed with the restaurant-friendly name "snow crab."

Commercial fishing is the most dangerous job in the United States. In 2008 the annual fatality rate among all U.S. fishermen was thirty-six times higher than for all U.S workers (128.9 and 3.6 deaths per 100,000 workers, respectively, according to the National Institute for Occupational Safety and Health). In the 1990s, the rate was even higher for Bering Sea crab fishermen. Between 1990 and 1999, seventy-three people died in the crab fishery, a number that translates to 768 annual fatalities per 100,000 full-time workers.

The gear is a major culprit: The crab are caught in rectangular metal traps, or pots, which are baited with herring and left to soak for up to two days at a time. Each pot can weigh 800 pounds and is launched into the ocean attached to a long line that connects the trap, which rests on the ocean floor, to a buoy on the surface. It's not unusual for a crewman—especially a newbie, or greenhorn—to be pulled overboard after getting an ankle or a loose piece of clothing wrapped up in a line.

When not in use, the pots are piled high atop slippery decks. Crewmen climb on the unstable stacks to tie down the pots and can easily fall several stories to the deck, or worse, into the ocean. The piled pots also diminish a boat's stability. Crab pots on deck make a ship top-heavy, which makes it roll more easily and right itself more slowly—if it rights itself at all. During the 1990s, twelve crab boats capsized and sank in the Bering, at least eight of them while traveling to or from the crab grounds with pots loaded high on deck.

Location also adds to the danger. The Pribilof Islands hug the 57th parallel, more than two hundred miles north of Dutch

Harbor and seven hundred miles west from the Coast Guard air station in Kodiak, one of two Alaskan bases equipped with HH-60 rescue helicopters. The Coast Guard's second air station is in Sitka, six hundred miles south in the Gulf of Alaska. Together, the two stations cover an area half the size of the continental United States. Even if the Coast Guard instantly got the call for vessel in distress or man overboard, it would take at least six hours for a helicopter to reach St. Paul from Kodiak. The HH-60 (also called the Jayhawk) is the Coast Guard's long-range helicopter, but it still wouldn't be able to make the trip without stopping to refuel at Cold Bay or Dutch Harbor, or some other Bering Sea outpost almost as far-flung as St. Paul. Six hours is too long in the Bering Sea. And so, from January through March, Coast Guard rescuers rotate through winter duty on St. Paul, sleeping, eating, and, often, looking for ways to pass the time in the barracks of the Coast Guard's LORAN station.

LORAN IS AN ACRONYM for "long range navigation." Like the Coast Guard's manned, high-frequency radio communication station, the LORAN facility was a relic of an earlier age. To most modern ocean-going vessels, it operated a technology that was as outdated as the sextant. The St. Paul LORAN station was built in the 1940s and was in 2008 one of about fifty remaining LORAN stations worldwide, each supporting a massive antenna that broadcasted a low-frequency radio signal hundreds, sometimes thousands, of miles in every direction. It takes at least three stations to provide the triangulated data needed to determine an exact position in the open ocean. The technology also requires a LORAN box, a piece of electronics that was ripped out of most vessels at least a decade ago.

St. Paul's LORAN station was the largest of six in Alaska,

the state's "master" station. Like all the LORAN stations in the United States, Alaska's facilities were run at considerable cost by the Coast Guard. The military justified the expense by arguing that LORAN was a backup to GPS. If the United States' entire system of orbiting satellites were shot out of the sky, LORAN would be the fallback. In the meantime, the remote, multimillion-dollar facilities were kept operating to serve the odd old-time fisherman who avoided dropping a few hundred dollars on a GPS in favor of antiquated electronics that served him just as well.

The staff of thirteen full-time Coast Guard personnel on St. Paul called themselves the "permanent party," and they didn't dwell on who was or wasn't using their broadcast. Their mission was simply to keep the one-megawatt signal always on air, day and night. The conditions sometimes made it difficult. In the summer, St. Paul sees highs in the low sixties. But in the winter, sub-zero temperatures are the norm. Thirty-five-mile-per-hour winds are nothing special and once or twice a season gusts in the sixties are virtually guaranteed. It's easy to get snowed in for days at a time. The LORAN crew was responsible for keeping the station and, most important, the 625-foot LORAN tower clear of snow. They had a full-size backhoe, some smaller snowmovers, and an eight-man snowcat to assist with the job.

The tower rose a few hundred yards behind the single-story building where the permanent party lived during the year-long assignment. St. Paul was a hardship post; each Coastie came alone—no spouses, no kids. For each month of the tour, a Coast Guard member earned an extra 2.5 days of vacation time and a $150 hardship bonus. The single best thing about the billet was that after a year at St. Paul, Coasties were virtually guaranteed their number-one choice for their next assignment. In the meantime, they lived like college students: working, sleeping, and eating at set mealtimes. The rooms were dorm-size, the furniture straight out of a 1990s-style freshman suite. Signs in the communal bathrooms reminded people to wipe the stainless steel sink after brushing their teeth.

There had been efforts to make the LORAN station a friendly place. There was a pool table and a foosball game, a TV room decorated with palm fronds known as the "Tundra Dome" where Coasties could pay 40 cents for a Coke or 89 cents for a Bud Light and watch a movie selected from one of dozens of new releases provided through the military's "morale" program. The adjacent building had a gym with treadmills and weights and a couple of old mountain bikes that staff might use to ride into town in warmer weather. The port's there, as well as a small museum, and the island's only store, which sells groceries and bathroom supplies and knock-off *Deadliest Catch* sweatshirts. You can buy an ATV there, or a bunk bed, or a $9 bag of Doritos. Sometimes the Coasties went there just to look, just for somewhere to go.

Despite their shared isolation, the LORAN staff usually didn't get to know many of St. Paul's full-time residents. In 2001, the commanding officer of the station was brutally beaten in his room with the butt of a gun, and then dragged outside the station and shot to death. The crime was the result of an apparent love-triangle—the murderer was the estranged husband of a local woman who was allegedly involved with the senior-level Coastie. The lurid details of the murder were among the very few facts new arrivals to the station might know about the place. The event didn't improve the already-cool relationship between the Coast Guard and the local community.

ZACHAROF. LESTENKOF. MERCULIEF. MELOVIDOV. Like in many of Alaska's native communities, the surnames on St. Paul are Russian, even though 85 percent of the island's residents have Aleutian ancestry. With a population of 450, St. Paul is the largest Aleut village in the state. Russian traders first "discovered" the Pribilof Islands in 1786. Aleut history holds that foreign sailors were led there by a native hunter. For centuries, the most striking thing about the place has been its fecund population of northern fur seals: Each summer, hundreds of thousands of the sweet-faced mammals gather on the islands' shores. The number used to be in the millions.

The Russians already had established settlements in Dutch Harbor and on Kodiak Island in the 1790s when they began forcibly relocating Aleut people from the Aleutian Chain north to St. Paul and the nearby island of St. George. On the Pribilofs, the Aleuts were forced to hunt and skin seals for the Chinese market. Over the years, many of the Aleut women married Russian men, and virtually all of the native Alaskans joined the Russian Orthodox religion, whose onion-domed churches remain the most distinctive buildings in St. Paul, St. George, and many other small Alaskan communities.

St. Paul residents still hunt fur seal. The "harvest" is for subsistence only, and is subject to strict government regulation. About sixteen hundred sub-adult males can be taken each summer. Only Aleuts can participate in the hunt, and only they can eat the meat. Not only it is illegal to sell the lean, bloody steaks, it's against the law for hunters to share the seal meat with outsiders, even over their own dinner tables.

Traditionally, Aleut men hunted fur seals with a harpoon, from a kayak. The animals rarely come ashore near their Aleutian Chain villages. In the Pribilofs, a more efficient method was, and still is, used: The animals are herded from the beach into a pen, much like cattle. Then, one hunter clubs the seal on the head to stun it, while a second hunter stabs the animal through the heart. The process was designed to avoid damaging the seals' valuable fur. Today, that fur is discarded—it, too, is illegal to sell, and both the equipment and skills that St. Paul hunters once used to dry and preserve the pelts have been lost.

Most of the money that comes into the St. Paul community comes through fishing. Aleut fishermen spend a good part of the summer trolling for halibut, a high-value white fish whose harvest is managed through quotas that are largely reserved for native communities. There are two fish processing plants in town: the Trident Seafoods factory on St. Paul Harbor, and the *Arctic Star,* a floating processor owned by Icicle Seafoods. In the warmer months, the island supports a small tourist trade of extreme bird watchers. More than 240 avian species have been spotted on the island, including some exotic Asian specimens.

In the winter there isn't any tourism and almost no activity in town. For many Coasties, the best thing about the place is the caribou (reindeer, technically) that were imported as an alternative food source when the seal population plummeted in the early 1900s. Today, a herd of about five hundred roams the island, which is more than the land can comfortably support. The local tribal council happily allowed the Coasties to buy a $50 tag and do some culling. They dressed and packaged the meat in an old metal storage container behind the LORAN station, then packed it in coolers for the flight back to Kodiak.

When hunting season was over, there was still hiking. Many of the "airdales" would pack snowshoes and trekking poles and walk along the ice at the edge of the beach. Inside, they played poker, Scrabble, Trivial Pursuit, and Xbox games. They tried to stay out of the way of the full-time LORAN staff.

The air crews' shifts were twenty-four hours on, twenty-four hours off, from 5:00 P.M. to 5:00 P.M. One crew was always ready to go should a call come in. In recent years, there'd been an average of half a dozen search and rescue cases each winter.

The crews flew training missions when they could. The Coast Guard has strict safety standards for training: a five-hundred-foot ceiling and two miles of visibility at takeoff, and maximum winds of 35 knots. When the weather was clear, they'd take familiarization flights around the islands or work on stick and rudder skills over the runway, practicing takeoffs and landings. There was no crash crew at the airport in St. Paul, and no second Coast Guard aircraft. Without the backup, the crew couldn't practice many of the drills that filled much of their time in Kodiak: simulating engine or systems failures, or practicing hoisting their rescue swimmer and rescue basket out of the ocean.

But like cops on a beat, the air crews could fly fisheries patrols. Sometimes, they'd use Coast Guard intel or a tip from the National Marine Fisheries Service to target a specific vessel suspected of fishing in a closed area or using illegal gear. More often, they were just checking up on ships from above, letting them know the Coast Guard was there if needed.

Occasionally, the crews conducted drills with a Coast Guard cutter. Earlier in the week, both Shawn Tripp's and Steve Bonn's helicopter crews had been scheduled to practice a maneuver known as HIFR (helicopter in-flight refueling) with the *Munro*, a 378-foot cutter on winter patrol in the Bering Sea. The ship is big enough to carry its own search and rescue aircraft, the French-made HH-65 Dolphin, which is stored in a snug hangar on the stern of the ship. The day Tripp and his crew were scheduled to train with the *Munro,* the weather was crummy. The next day, though, was clear and calm, a perfect training day for Bonn's helicopter crew.

They came into a hover forty feet over the *Munro*'s deck. Thirty-year-old lieutenant Brian McLaughlin was at the controls; Bonn, a former Army pilot from Northampton, Pennsylvania, who'd been in Alaska for four years, was his copilot. The ship's seamen had laid out a fuel hose in a wide S-like shape. The flight mechanic in the rear of the Jayhawk lowered the helicopter's external hoist line—a steel cable one-fifth of an inch in diameter with a talon hook on the end—and the ship's crew attached their hose to the hook. The line was raised, and the pilots backed the bird off the side of the ship. The flight mechanic inserted the gas nozzle into their internal fuel tank and began refueling. The seas were calm and the winds were low, and Bonn could see the shadow of the helicopter on the cutter's deck. Each member of the crew had studied every step in an HIFR, but neither of the pilots had ever actually practiced the skill. It was satisfying to drill with the *Munro* and they were grateful for the opportunity to do it.

"*Alaska Ranger*, this is COMMSTA." It was 2:49 A.M., just a couple minutes after the fishing trawler's initial Mayday call. Inside the *Alaska Ranger*'s wheelhouse, First Mate David Silveira was handling communications with the Coast Guard, while Captain Pete Jacobsen consulted with the *Alaska Ranger*'s engineers. Meanwhile, the Japanese crew was sitting in a circle on the floor near the back of the wheelhouse, smoking cigarettes.

"Understand not able to keep up with the flooding and all your crew members have their survival suits on and are standing by at this time, over," watchstander David Seidl confirmed with the mate.

"That's a roger," Silveira answered.

"*Alaska Ranger,* this is COMMSTA. Request that you turn your EPIRB [emergency position-indicating radio beacon] on immediately and keep it with you, over."

"Roger that. Roger that."

"*Alaska Ranger,* this is COMMSTA. Request to know what

type of survival gear and flotation devices you have on board, over."

Silveira replied that the boat was equipped with immersion suits for everyone on board, and with three twenty-man life rafts.

"*Alaska Ranger,* this is COMMSTA. Understand, twenty-man life rafts. . . . Request to know if you are able to give me on-scene weather, over?"

"On-scene weather is northwest winds about three, five knots, northwest about thirty-five knots," Silveira reported to the Coastie: gale-force winds of just over 40 miles per hour. "We've lost our steering, uh, we don't have any steering. We haven't lost power yet, the engines are still on."

At 2:55 A.M. COMMSTA Kodiak issued the first Urgent Marine Broadcast alerting all Bering Sea traffic about the foundering ship. Like the standardized maritime distress call—the thrice repeated "Mayday," which comes from the French *m'aidez,* or "help me"—at-sea alerts begin with an anglicized version of a French word, *panne,* or "breakdown."

"Pan, pan. Pan, pan. Pan, pan. Hello all stations. This is United States Coast Guard, Kodiak, Alaska, Communications Station. United States Coast Guard, Kodiak, Alaska, Communications Station. The factory trawler *Alaska Ranger* is taking on water in position 5, 3, 5, 3.4 north, 1, 6, 9, 5, 8.4 west. There are forty-seven persons on board, and it is one hundred and eighty-four feet with black hull and white trim. All vessels in the vicinity are requested to retain a sharp lookout, assist if possible, and report all sightings to the United States Coast Guard. United States Coast Guard, Kodiak, Alaska Communications Station. Out."

The officers in the *Ranger*'s own wheelhouse heard the chilling announcement over HF channel 2182. It was unlikely, they knew, that an unknown Good Samaritan vessel was close by.

The most likely "Good Sams" in the vicinity were the other Fishing Company of Alaska trawlers. The closest was the *Alaska Warrior,* which was more than forty miles away.

"*Alaska Ranger,* this is COMMSTA Kodiak. At this time, we'd like to put you on a zero-five minute communication schedule," Seidl radioed at 3:00 A.M. "We'll contact you every five minutes for updates on your status, over."

"Roger," First Mate David Silveira answered.

A few minutes later, Seidl called to ask for an update on the flooding.

"Well, it's over our, uh, we call it the ramp room. Our rudder room was flooding, coming up the ramp room. We've shut the watertight doors," Silveira reported. "We got out of the area." Heavy static distorted the second half of the transmission, but it sounded like Silveira was saying that the *Alaska Ranger*'s chief engineer was recommending they abandon ship.

"*Alaska Ranger,* this is COMMSTA," Seidl radioed back to the boat. "Understand above the rudder room and to your ramp room. You shut the watertight doors, got out of the area, and donned your survival suits, over."

"That's a roger."

"*Alaska Ranger,* this is COMMSTA. Nothing further. Talk to you in five minutes, over."

"Roger," Silveira answered.

OUTSIDE ON THE DECK, SEVERAL PROCESSORS were leaned up against the rail. Among them was David Hull, a twenty-six-year-old from Seattle who had been working on the *Ranger* on and off for the past five years. David was well known as the ship's health nut. He kept a blender in his room, along with supplies to make smoothies, and a large collection of vitamins.

It was after the muster, and things seemed pretty quiet. David looked around. He was thinking about the valuables he'd left behind in his room. He felt like things were calm enough that no one would notice if he was gone for a few minutes. He ran as fast as he could in his suit, down two flights to his bunk room, where he grabbed his laptop bag and started stuffing a bunch of vitamins inside. Before long, boatswain Chris Cossich—who was in charge of David's muster group—was at the door. "Heh! Get out of here," he yelled. Chris was furious.

"Fuck you! You're fired!" he yelled at David when the two men got safely back up on deck. "What the hell were you thinking? You are off this boat!" David felt terrible. He realized he'd made a stupid move. He tried to apologize, but Chris wasn't having it. It was pretty obvious to everyone who heard the commotion that descending into a sinking ship to get a computer bag was about as smart as running into a burning building.

PILOT SHAWN TRIPP WAS TIRED. He had landed back in St. Paul just a few hours before, after a four-and-a-half-hour medevac flight to Dutch Harbor and back. A forty-nine-year-old man in Dutch had needed an emergency blood transfusion. The simplest thing would have been to put him on a plane directly to Anchorage, which has the state's best-equipped hospital. But Dutch Harbor's tiny airport was completely socked in. A plane couldn't land. But a helicopter could. The helo in St. Paul—Coasties prefer the shortened term to "chopper" or "copter"—was the closest available aircraft.

In the Lower 48, the Coast Guard rarely performs medevacs, except in civic emergencies, like Hurricane Katrina, when the Coast Guard transferred more than nine thousand patients out of battered New Orleans hospitals and nursing homes. An additional twenty-four thousand civilians were rescued by the Coast Guard from rooftops, floating debris, and even tree branches in the days following the storm. The Katrina tragedy was a shining moment for the Coast Guard. It showcased the strength and flexibility of the service's real-time planning and response capabilities, and allowed the Coast Guard to demonstrate its willingness to step up and deal with problems that technically fall under other agencies' purviews. In Alaska, that sort of stepping up happens every day.

Alaska is bigger than four Californias put together—and has a population of just 650,000 people—less than the city of Columbus, Ohio. It's unsurprising, then, that so many of Alaska's communities are cut off from the rest of the world. Only in the heart of the state, branching out from Anchorage, where half the population lives, do maintained roads connect communities on a year-round basis. In many areas the only way to move between towns is by boat or plane. Many remote towns don't have a real hospital. And even those that do often aren't equipped to handle high-risk procedures. Or even low-risk ones: There's only been one baby born on St. Paul Island in twenty years, a little girl who arrived prematurely. At eight months, expectant mothers are ordered to Anchorage.

The isolation means that medevacs are high on the Coast Guard's list of calls. In the summertime, it's cruise-ship passengers from the massive boats that trace the Kenai Peninsula, or the smaller vessels that come into Kodiak and very occasionally visit the Aleutian Chain. Hunters, hikers, four-wheelers, and snowmobilers routinely get themselves into trouble in Alaska's unforgiving mountains. The massive shipping fleet whose routes ply the Bering Sea are regular customers as well. It isn't unusual for a rescue crew to be sent out beyond Adak, a former military base two thousand miles west of Kodiak, to lift an injured worker off a four- or five-hundred-foot container ship.

This medevac had been fairly routine, even though the bad weather and treacherous flying left Tripp wired as he arrived back at the LORAN station. The night vision goggles he and his crew wore in the helo made the Bering look like the opening credits of *Star Trek*—the snowflakes were like a universe of stars hurtling toward him at light speed. Tripp had sixteen hours left on his shift. He'd pass a couple hours with Call of Duty. He'd had an ongoing competition with pilot Steve Bonn. They were well matched in the game: equally terrible.

The phone rang a couple minutes before 3:00 A.M., just after the men had finished a final face-off. Tripp figured it was the on-duty officer at the operations center in Kodiak, Todd Troup, calling to point out some mistake Tripp had made in his medevac paperwork. It was the ops center, all right, but Troup wasn't concerned about paperwork. A fishing trawler was taking on water, some two hundred miles south of the island. The 60 Jayhawk in the St. Paul hangar was the Coast Guard's closest asset.

Tripp did the calculations. The ship was at least an hour-and-a-half flight away. His crew had already had four and a half hours of flying time. A crew is "bagged," or grounded, after six hours in the air. Of course, if they were in the middle of a rescue, they would keep going until it was over, but in this case, Tripp's crew would have close to six hours on them before they even reached the troubled ship. Tripp knew it didn't make sense for his crew to respond, and Troup had reached the same conclusion.

Tripp held out the phone for Bonn: "It's for you."

Minutes later, Bonn was knocking on doors to wake the rest of his crew: pilot Brian McLaughlin, flight mechanic Rob DeBolt, and rescue swimmer O'Brien Starr-Hollow. Though at thirty-nine, Bonn was older than McLaughlin and had more years of flying experience, McLaughlin outranked him in the Coast Guard.

The younger pilot was tall, six foot four, and lanky, with pale skin and sharp features. He had enrolled in the Coast Guard Academy right out of high school in Hanson, Massachusetts. As a kid, McLaughlin had been in the Civil Air Patrol, a sort of military Boy Scouts. In eighth grade, he attended a Civil Air Patrol camp that included a visit to the Coast Guard Air Station on Cape Cod. He left with a new goal in life: to become a Coast Guard pilot. He decided his best route was the Coast Guard Academy. It was one of the most difficult schools to get into in the country. The tuition was free, and the academic standards were high. McLaughlin was a trumpet player, a self-declared band geek, and the Academy had a band, of course. The school wasn't far from his home, just a couple of hours away in New London, Connecticut. He applied, and got in.

It wasn't a typical college experience. McLaughlin reported to New London in July 1995, for Swab Summer, a six-week basic training program for new cadets. There, he learned to sprint through obstacle courses, to fire an M-16, and to recite the Academy's cadet mission statement: "To graduate young men and women with sound bodies, stout hearts, and alert minds, with a liking for the sea and its lore, and with that high sense of honor, loyalty, and obedience which goes with trained initiative and leadership; well grounded in seamanship, the sciences, and amenities, and strong in the resolve to be worthy of the traditions of commissioned officers in the United States Coast Guard in the service of their country and humanity."

Along with about 240 other first-year cadets, McLaughlin spent his freshman year in New London walking silently in the hallways and greeting any upperclassman he encountered by name. In the embarrassing instances when he couldn't recall a name, he had to greet the older student with "sir" or "ma'am." Glancing down at the name tags embroidered on the upperclassmen's uniforms was forbidden—new cadets were required to keep their chins up and their gaze straight ahead at all times. At meals, the freshmen sat together under strict silence in the cafeteria—the "ward room" they called it, just like the officer's dining room on a ship. McLaughlin was instructed to sit straight up, with a fist's distance between his back and the back of his chair. He was taught to "square his meals," by raising a fork straight up from the plate to a few inches in front of his face before bringing it forward into his mouth.

The students were forbidden from closing their dorm-room door any time there was a cadet of a different year, or of the opposite sex, in their room. Freshmen were allowed to date only within their own class, and upperclassmen could date only one year in either direction (no senior/sophomore relationships allowed). Romance rules weren't relevant to McLaughlin, who had started dating Amy Lundrigan, from the neighboring town of Whitman, at the end of high school. They went together to their senior prom and decided they'd stay together when McLaughlin left for the Academy that summer.

There were no phones in the dorm rooms, but the couple scheduled pay phone calls. Amy would sometimes drive down on weekends and stay with a friend nearby. McLaughlin could spend weekends with her, but he had to be back for curfew: 10:00 P.M. on Saturday and 6:00 P.M. on Sunday. They stayed together for four years and got married a few months after McLaughlin graduated from the Academy. After the requisite year and a half afloat on the 270-foot Coast Guard cutter *Tahoma* (he and Amy called it the *Tahoma Neverhoma*), McLaughlin was accepted to flight school in Pensacola, Florida.

All Academy graduates are committed to at least five years of Coast Guard service and they are an elite group among the ranks. There are currently just over forty thousand active-duty

Coast Guard members. Of those, 11 percent are Academy graduates. Among pilots, the number is close to 39 percent. Upon graduation, a twenty-one-year-old Coastie is already a junior-grade lieutenant, outranking enlisted officers who've been in the Coast Guard for decades. McLaughlin got his wings at twenty-four and became an aircraft commander at twenty-six. He'd been stationed in Kodiak since July 2006. It was his second assignment as a pilot after Clearwater, Florida.

Clearwater was an excellent post for new pilots; conditions were almost always good for flying, which made it easy to rack up a lot of hours. The air station had diverse missions: migrant operations, hurricane response, search and rescue, and, always, recreational boaters who found creative ways of getting into trouble. There were also a high number of false alarms. "Condo Commandos" was the term the pilots used to describe the overzealous Floridians who called in flare sightings from the balconies of their beachside second homes. What they'd usually seen was an odd firework. Still, the Coast Guard dutifully went out to search. There was a lot of flying, but there often wasn't a lot of true action.

Alaska was a different story. There weren't as many search and rescue cases. Sometimes a week would go by without an incident. More typically, the ops center at Kodiak would get three or four calls a week. But in Alaska there were few false alarms and few small, no-big-deal kind of cases. Almost every time McLaughlin flew, he was dealing with long distances, icy conditions, turbulence, and high seas. Down in Clearwater, ten-foot seas usually meant a tropical storm was coming in. In Kodiak, ten-foot seas were the norm. In less than two years in Kodiak, McLaughlin had already been on two major rescues that involved pulling multiple people from the ocean, as well as a handful of more typical medevacs and missing hunter calls.

After being woken up by Bonn, he changed into his thermal

underwear and orange dry suit. He'd gone to bed just a couple hours before, around 1:00 A.M. He'd been awake when Shawn Tripp returned from his medevac, and he had heard Tripp and his copilot talk about the "snotty" weather farther south. Too bad for you, McLaughlin had thought to himself. Those guys were on duty until the next afternoon.

Now, things had changed.

It didn't take McLaughlin long to pull together his gear. Coast Guard rescuers are encouraged to compile their own custom survival kits for the emergency conditions they may face in the region. McLaughlin carried a hunting knife, a compass, two space blankets, a lighter, waterproof matches, and a snapshot of Amy with their two kids.

Their daughter, Sagan, had been less than a year old when they made the trip from Florida to Kodiak. They spent the whole summer at it, crossing the continental United States and then making their way up the Alaska Highway in a twenty-three-foot RV they'd bought before leaving Florida. It was just them, the baby, and their two Australian Shepherd mixes, Sadie and Roxie.

McLaughlin had talked to his wife on Saturday evening. Their son, Cole, had been born just a month before, and McLaughlin's mom had come from Massachusetts to help out. The women were planning an Easter dinner and getting Sagan's basket ready for the morning. Amy told her husband about how there was an egg hunt, and pictures with the Easter Bunny, at the air station that day. There were lots of activities for families at the base, and lots of families with little kids. Amy had worked as a massage therapist in Florida, but she'd been a full-time mom since moving to Kodiak. She'd bring her daughter swimming at the air station's indoor pool and to the small aquarium run by the National Marine Fisheries Service, where Sagan loved to stick her tiny hands in the freezing touch tank and run her fingers over the prickly starfish and tissuelike sea anemones.

A couple of nights a week, McLaughlin would be on duty, sleeping at the base. It was the two-week deployments to the remote outposts that were hard, though. Each season was a new place: St. Paul in the winter, Cordova in the summer, Cold Bay in the fall. Amy didn't worry too much. Brian's father was a Massachusetts state trooper, and her mother-in-law had taught her that "You can't worry every time they walk out the door. You'll drive yourself crazy if you do."

Amy called the deployments "man camp." Eat, sleep, movies, video games. She didn't feel so bad for her husband. He'd already been gone a week and half. By the end of the month he'd be home.

The temperature outside was –11°F with windchill as the men loaded into the SUVs and headed toward the hangar. It was squalling, with 30-knot winds, and the two vehicles backed off from each other when they reached several large snowdrifts that had blown over the road. One by one they gunned it, barreling through the heavy drifts and spinning and sliding the rest of the way down to the airport.

As the pilots got their gear in order, flight mechanic Rob DeBolt helped the line crew move the 14,500-pound helo from the hangar onto the icy tarmac. DeBolt was twenty-eight years old and had grown up in Walla Walla, Washington. He'd been enlisted in the Coast Guard for eight years. He hadn't seen that many cases, though. Just a couple of easy medevacs. And lots of training. The mechanics secured the Jayhawk's front wheel to a tow bar, and then the aircraft was tugged out of the shelter with a golf-cart-size vehicle known as a mule.

On the drive to the hangar, the pilots had told the rest of the crew what little they knew of the case. The boat was big, almost two hundred feet, and the ops center had said there were forty-seven people on board. They should bring a mass casualty raft, McLaughlin thought. It was 100 pounds and would take up quite a bit of room in the cabin. But it could hold twenty people. They'd also bring a dewatering pump, which they could drop to the deck of the fishing boat with their hoist cable. The pump was also heavy: 88 pounds, protected by a hard plastic case.

From nose to tail, the Sikorsky helicopter is sixty-five feet long. The cabin, though, isn't any bigger than the inside of a typical SUV. The extra gear would take up considerable space. McLaughlin thought about his earlier rescues. The biggest one had been five people in the cabin, in addition to the aircrew. That had been damned crowded. They'd bring the extra equipment, though. If they needed to, they could ditch it in the ocean.

McLaughlin climbed into the helicopter and punched the *Alaska Ranger*'s coordinates into the aircraft's computer. The ship was 197 miles south of St. Paul. There was a tailwind. Still, they'd load the aircraft with the maximum fuel allowance. When the crew got to the hangar, the helicopter was already gassed up with 5,000 pounds of jet fuel (736 gallons), the normal load for a take-off from St. Paul. The crew added another 1,200 pounds—the aircraft's "max gas"—which would give them an extra hour of flying time. McLaughlin was in the left seat, Bonn in the right. The Jayhawk can be flown from either seat, though the flying pilot usually sits to the right. If needed, McLaughlin could jump in at any time to take control of the aircraft from Bonn.

It was just before 4:00 A.M., with sunrise more than five hours away, when the crew slammed shut the helo's doors. DeBolt and Starr-Hollow buckled themselves into jump seats in the back, and the helicopter lifted off into the black night.

Chapter Four

Best Speed

The water was still, the surface smooth. The Coast Guard had a term for the conditions, FAC: flat-ass calm. When the ocean was this way, it felt more like floating in a lake than in the Bering Sea. A very cold lake, thought Operations Boss Jimmy Terrell. Slabs of broken ice were scattered across the glassy ocean. It was Saturday afternoon, and though the sun was close and bright on the horizon, the temperature hadn't nudged above freezing all week. Terrell had been stationed on board the cutter *Munro* for a year and a half, and he'd never seen conditions quite like it before.

The 378-foot ship was patrolling near the Arctic ice edge, up near the Pribilof Islands. They'd turned off the engines. It was unheard-of to drift in the Bering, but the ice edge had pushed far south this season and the wind had been from the north for weeks. Deep in the ship the crew could still feel the hum of the generators, but on deck everything was still. It was quiet, like floating near shore in a sheltered cove when the breeze is blowing out to sea.

Terrell was thirty, from El Paso, Texas. Ever since his graduation from the Coast Guard Academy seven years earlier he'd been following Craig Lloyd, a twenty-four-year Coastie and the *Munro*'s captain. They'd worked together on the *Mellon*, a 378-foot cutter out of Seattle. Then, both men transferred to Alameda, California, where Lloyd served as the chief of cutter forces for the entire West Coast. Terrell was his admin guy. The placements were just a coincidence, but the two joked that Terrell had become Captain Lloyd's permanent lackey.

By August 2006, both had moved to the *Munro*. Under Captain Lloyd's command, the ship had gained the reputation of being one of the best run boats in the fleet. It was a good thing, because the Bering Sea wasn't exactly the most popular place to spend a couple years working on board a Coast Guard cutter. San Diego, the Caribbean, the Gulf of Mexico—any of them sounded a whole lot better than winter in Alaska.

The *Munro*'s crew numbered 160. Under Captain Lloyd was Executive Officer Mike Gatlin ("XO" the crew called him), and Operations Boss Terrell ("Ops"), who was responsible for the ship's operational strategy, and served as a mentor to the *Munro*'s dozen junior officers, most of them recent graduates of the Academy. The majority of the crew was enlisted men and women, many of them on their first deployment, some of them only eighteen or nineteen years old and just weeks out of boot camp in Cape May, New Jersey.

Like Terrell, the typical crewman was serving a two-year billet on board the *Munro*. The previous fall, the ship's home port moved permanently north from Alameda, California, to

Kodiak. Most of the crew kept homes in town, where they'd stay for the months between deployments while the ship remained tied up at the base's pier. The *Munro* had set sail for a two-month deployment less than two weeks before. On the day the ship left port, wives, husbands, boyfriends, and girlfriends clustered on the broad, wooden pier, waving good-bye as the huge white ship backed away from the dock and slowly vanished down the mouth of Womens Bay.

Reveille was at 6:45 A.M. The days were tightly scheduled: meals, training, cleaning, more training. The *Munro*'s mission included law enforcement, which involved at-sea boardings of fishing vessels. The crew was constantly conducting drills: fighting mock fires, controlling hypothetical flooding, or rescuing plastic dummies flung over the side of the ship. The drills were timed, and the minutes that the victim had been overboard were piped over the shipwide intercom system. When the drill was complete, the crew were told if they passed or failed. More than seven minutes meant the dummy was near death when it was hauled back aboard—and that the drill would likely be repeated in coming days. When the weather was cooperating, the *Munro*'s crew assisted the helicopter team deployed aboard the ship with its own flight training regimen.

The cutter often felt like a floating technical college: On the bridge, young Coasties learned to drive the ship, to navigate, and to read the charts and constant streams of weather data that kept the boat safe in one of the world's most violent oceans. Five decks down, in the engine room, others learned the ship's mechanics: how to spot the earliest signs of an engine problem or oil leak, how to manage the ship's tanks and optimize its fuel use. Meanwhile, other seamen learned to cook, clean, and dispose of the ship's trash. Senior enlisted men apprenticed young officers on managing the ship's supplies and finances. Everyone had a job to do.

The *Munro*'s primary mission, though, the duty that came before all others, was search and rescue.

Captain Lloyd had spent a total of six years in Alaska and the *Munro* was his third Alaskan ship. He knew the Bering and its traffic well, and he'd chosen the ship's ice-edge position out of concern for the fishing fleet. If he kept his ship north of the fleet, he could sprint downwind in case of an emergency. He'd make better speed if he wasn't fighting the seas. And the strategy had another advantage as well—pounding into the swells in cold conditions can cause sea spray to turn to ice on the bow, the rails, and any other exposed area of the ship. Fishermen call it "making ice," and it can have catastrophic effects on a boat's stability. Terrell had experienced it several times back on the *Mellon,* where a senior officer had taught new crewmen about the danger of ice by warning that a half-inch of buildup around a 378's pilothouse has an effect equal to parking a Ford F-150 on the house's roof. When more than a few centimeters of ice built up, crews were sent out on deck with wooden mallets and baseball bats to bang it off, which was a dangerous undertaking in itself. (On the *Mellon,* those deicing details were named after professional baseball teams.)

By late March, the crab season was ending. The ice edge was nudging so far down that the opilio boats had been pushed out of some of their favorite northern fishing grounds. There were a lot of ships farther south, though—trawlers, mostly. Many of them were factory boats with large crews of twenty, forty, even fifty men headed to fishing grounds hundreds of miles from the closest port. The Coasties called them draggers, and they weren't always the most popular boats to deal with. The Coast Guard didn't need a specific reason to board a fishing vessel at sea, but they often targeted boats that had avoided the Coast Guard's voluntary inspections in Dutch Harbor or ended their dockside visit with holes in their safety checklists.

The *Munro* would radio the fishing vessel that they were coming, then send a half dozen crew in one of the cutter's two orange, rubber Zodiacs, which were stored in cradles on either side of the ship. The boardings usually kept the boat away from fishing—and profit—for a couple hours. Not too many fishermen were thrilled to see a Coast Guard cutter on the horizon.

THE VAST MAJORITY OF FISHING BOATS are classified as "uninspected vessels" and aren't required to be classed and load-lined like cargo carriers and passenger ships. Nongovernmental classification societies determine construction standards for most large boats and periodically examine them to be sure they remain up to code. Among those standards is the load line test, which ensures a boat has a good watertight envelope. Go to any large marina, and you'll notice physical load lines—painted stripes around a ship's hull right above the waterline. A load line test establishes how low a ship can safely sit in the water, how heavily it can be loaded, and how the weight can be safety distributed—as well as that different compartments of a large boat are watertight from one another. Class standards, meanwhile, focus on the upkeep of a ship's engines and electrical systems. With just a few exceptions, though, fishing vessels are immune from those standards.

Coast Guard officers in Dutch Harbor do regularly examine most of the boats in the Bering Sea fishing fleet. Unlike the inspections for cargo ships and passenger vessels, however, the fishing boat examinations are technically voluntary and don't focus on the seaworthiness of the ships. What they do focus on is safety equipment. Maybe the Coast Guard can't prevent a boat from sinking, but at least they can give the fishermen a decent chance at survival if it does.

At the request of vessel owners, once every two years Coast Guard fishing vessel examiners in Dutch Harbor board most boats in the fleet and fill out a safety equipment checklist. Ships are required to have life rafts sufficient to carry everyone on board, as well as a survival suit for each crew member. (The spongy coveralls are wide-legged and force the wearers to shuffle when they walk. They're often called "Gumby" suits, after the stop-motion clay character who walks the same way.) During each biennial inspection, the bright red suits are pulled from their individual bags and spread on the deck of the ship to be examined for tears and wear and to see if the zippers are in working order. The Coasties check if the ship has emergency flares, fire extinguishers, and an emergency position-indicating radio beacon (EPIRB).

Every boat should have this device, which can cost more than $1,000 and which sends a satellite signal with the ship's location and identification number when activated. Most are properly installed on an outside wall of the wheelhouse. Dead batteries are common, though. New Coast Guard fishing vessel examiners—especially those who are used to working with inspected vessels—are often surprised by how little attention some fishing boats pay to safety. Those who've been around for a while know things are a lot better than they used to be.

Before the early 1990s, even the most basic safety equipment was often absent on Alaskan fishing boats. Ships regularly went to sea with no life rafts, no survival suits—no way of letting anyone know when the worst was about to happen. Then, as today, many of the hired crew were novices, men who had little experience with the sea or with judging a boat's seaworthiness, certainly with no firsthand experience of the particular hazards of Alaska, with its unforgiving cold and freezing seas.

Peter Barry was one of those young men. It was the summer

of 1985 and he was nineteen years old, looking for an adventure out West before returning to his junior year at Yale. Peter was tall, six foot three, and slender, with a thick brush of blond bangs that swooped across his forehead. He spent a number of weeks working on the "slime line" in one of the Kodiak canneries. When a strike erupted among the mostly Filipino workforce in midsummer, Peter didn't want to cross the picket line. Instead, he started walking the docks and soon met Jerald Bouchard, captain of the *Western Sea,* a fifty-eight-foot wooden purse seiner that sailed with a five-man crew and had been fishing for salmon in northwestern waters since before the end of World War I.

Bouchard had just lost a deckhand. He offered Peter the job.

A few weeks later, another boat found Peter's body. He was floating facedown in the water, three miles offshore on the backside of Kodiak Island. He was dressed only in jeans and an old-fashioned Mae West–style life preserver.

The *Western Sea,* it turned out, had no life raft and no survival suits. There was no Mayday call. The first clue anyone had of a problem on the seventy-year-old ship was the discovery of the body, which was identified through a letter found in the pants pocket. It was from Peter's college girlfriend.

Despite an extensive search by the local Coast Guard, all that was found of the salmon boat was odd debris, a piece of the wheelhouse, and a life ring printed with the ship's name.

Several weeks later, two more bodies were found. One of them was that of a twenty-five-year-old crewman from Washington State, also a summer worker. The other was that of Captain Jerald Bouchard, whose decomposed remains were spotted floating in the ocean and recovered by crew of the cutter *Munro,* which was on patrol near Kodiak at the time. A toxicology exam revealed that the captain had cocaine in his system when he died.

* * *

Peter Barry's story wasn't that different from tragedies that were happening every year. It was a predictable formula: a captain with minimal regard for safety; an inexperienced crew member; and an old, undermaintained boat, often fishing in waters more dangerous than its size or condition warranted. Barry was one of 102 people killed on fishing boats in U.S. waters that year.

But Peter Barry was different from all those other fishermen. He was an Ivy League student. His father, Robert Barry, was a senior diplomat in the U.S. State Department. And his mother, Peggy, was a woman who would not allow her son's death to be dismissed as just another accident—nor tolerate the idea that the tragedy was either bad luck or God's will, as so many people would say about boats and men lost at sea.

The Barrys heard again and again that fishing was a dangerous business, that their son's death was a travesty, but that there was nothing to be done. They were told a ship at sea is at the mercy of the elements, that these things happen all the time, and that everyone who gets on a boat up north should know what they're getting into—that their fate is out of their hands. Fate? Peggy Barry thought. This boat had no life raft. It had no survival suits. This wasn't fate. A man can't survive in those temperatures for even an hour. The way she saw it, her son's death was criminal.

The Barrys began a campaign to mobilize elected officials, safety professionals, and the families of other lost fishermen in support of federal legislation to improve safety in the fishing fleet. Safety advocates had pushed for similar legislation in the past with no success. The fishing industry lobby was just too powerful, contributing hundreds of thousands of dollars each year to congressmen from Washington State and Alaska, most of whom remained opposed to increased regulation of commercial fishing.

The data were on the side of safety: In the mid-1980s, more than one hundred people were dying in commercial fishing accidents each year. Meanwhile, fishing vessels were the single major category of boat whose seaworthiness was not under government purview. It was clear that increased federal regulation had made a difference for other vessels. Small ferries and tourist boats, for example, had been unregulated until the 1950s, when a string of highly publicized disasters led Congress to act. Within a few years, the average annual fatality rate for that class of passenger boat had fallen from twenty-nine to five. Back then, boat owners had argued that the higher cost of meeting safety requirements would put them out of business (of course, fifty years later, we suffer no lack of duck boats and sunset cruises). Fishing vessel owners had long made the same financial argument. This time, though, the timing was right for safety advocates.

By the mid-1980s, the fishing industry was in an insurance crisis: High casualties had driven up insurance prices so much that half the ships in the Alaskan fleet could no longer afford their annual premiums. Fishing vessel owners wanted Congress to cap the amount injured fishermen (or family members of the deceased) could be awarded in a court case. The American Trial Lawyers Association was against the cap, which would limit their profits right along with awards for fishermen and their families. Both sides had lobbyists lined up to fight to the death.

The conflict created an opening for safety advocates like Peggy Barry, who helped to develop a safety bill that included an insurance cap (a sweetener to gain the support of boat owners). The trial lawyers shot that bill down. But Barry and her supporters didn't give up. They developed another bill; this one focused exclusively on safety. The Commercial Fishing Industry Vessel Safety Act passed on September 9, 1988.

Because of Peter Barry, every commercial fishing vessel in the United States today is required to carry safety equipment: life rafts, survival suits (for ships operating in cold water), signal flares, fire extinguishers, and a registered EPIRB. Emergency alarms and bilge pumps are mandatory on all ships, and crew are required by law to conduct regular safety drills. The results have been striking. Between 1990 and 2006, the annual fatality rate among commercial fishermen across the United States fell by 51 percent. Boats were still sinking—but men were surviving.

And yet the number of ships going down had barely changed at all. In the average year, 119 U.S. fishing boats are lost. That's one boat gone every three days. Most years, those losses are responsible for significantly more than half of the total fishing deaths (falls overboard and equipment-related accidents account for most of the rest). It was a source of frustration for some in the Coast Guard—among them fishing vessel examiners Chris Woodley and Charlie Medlicott.

Both were longtime Coast Guard men. Woodley had spent most of his twenty-year career moving back and forth between Alaska and Washington State. The Coast Guard's operating philosophy holds that diversity of geography and job experience is good, that moving constantly broadens a Coastie's experience and makes that person more valuable at their next post. But Woodley had grown up in Alaska. He went to graduate school in Seattle and wrote his master's thesis on impediments to safety improvements in the fishing industry. Even when he was based in Washington, Woodley spent a lot of time in Dutch Harbor, traveling north to help local inspectors like Charlie Medlicott at the start of the busiest fishing seasons. Most years, Woodley watched the Super Bowl with a bunch of fishermen at one of the town's bars.

Woodley and Medlicott had seen so many boats go down. Sometimes the sinkings were mysteries. The ship seemed as good as any other in the fleet; it had a competent captain, an experienced crew, a seemingly stable construction—and still it disappeared deep in the Bering. More often, boat losses fit a pattern. Especially with the crab boats. The damned things were so often overloaded by captains who didn't read their own stability booklet (loading guidelines prepared by a marine architect) or maybe just didn't care what it said. It was a recipe for disaster: a boat weighed down with crab, empty pots stacked high on deck—much higher than the architect had ever intended—and bad weather. The boat rolled, capsized, and sank.

By the late 1990s, Woodley and Medlicott had had enough.

Under U.S. Code, the Coast Guard has the authority to board any U.S.-flagged vessel, examine it, and—if the boat poses a threat to the environment, to commerce, or to human life—prevent that ship from leaving port. It was an authority that was hardly ever acted on in the Coast Guard. Mostly, the Coasties didn't want to piss people off. The other services call them "Boy Scouts with Boats." Most Army or Navy guys mean it as a jab, but many in the Coast Guard are Boy Scout types, and proud of it. These are men and women, after all, who at a young age decided they wanted nothing more than to become professional rescuers. They wanted stability and structure, the honor of the military, but maybe didn't want to go to war. They wanted to be the good guys. Why would they go out of their way to board boats against the owners' will when it wasn't really part of their job? When the law on the issue was anything but clear?

How about to save some lives? Woodley thought.

In October 1999 he and Charlie Medlicott started their own experiment, boarding every Bering Sea crab boat before it left port. A shocking number were overloaded. And it was tough for captains to argue when the inspectors pointed out a discrepancy between the number of pots on deck and the number authorized by the ship's own stability booklet. The captains pulled pots off, left them stacked at the dock, and returned to port with their boats intact.

It didn't take long to see a difference. By January 2005, there hadn't been a Bering Sea crab boat lost in five years. Charlie Medlicott was the law, but he'd been around long enough to be respected, even liked, by a lot of local fishermen. He was in his mid-forties, with ruddy cheeks, wild eyes, and a bow-legged stride. Charlie knew the fleet, knew the boats' histories and challenges. He wasn't above talking out a problem over a beer in the bar.

Charlie was different from many of the Coasties Dutch Harbor fishermen encountered, often young officers who'd never been in Alaska before they got stuck with this assignment, and who'd never be back after their year-long hardship post was up (like St. Paul, the assignment to Dutch was one the Coasties reported to without their families, for only a year at a time). Charlie, on the other hand, loved Dutch Harbor. He was fascinated by the place from the first time he landed there—the weather, the characters, all the big boats. He was a boat guy, after all. He'd served for years in the Coast Guard, worked at the small boat station in Juneau, and on the *Liberty*, a 110-foot patrol boat. He got out after that. He fished for a couple of years and then worked for a company selling marine supplies and safety equipment, packing life rafts. In 1993 he got back in the Coast Guard as a civilian employee. Since then, he'd been bouncing back and forth between Anchorage and Dutch Harbor. He knew a lot of people there; he was part of the community.

Charlie wasn't coy when he ran into Gary Edwards at the hotel bar at the Grand Aleutian one evening in January 2005.

"I'll be down tomorrow morning to check your boat," he told the captain of the crabber *Big Valley,* which was scheduled to leave Dutch Harbor to fish for opilio up near St. Paul Island. Gary was well-known around Dutch. He was probably the only guy in rural Alaska who regularly dressed in a tweed jacket and a beret and kept a Buddha shrine on his boat, complete with burning incense. In two previous instances, both men knew, the *Big Valley* had been overloaded when the Coasties came down to the dock, and Gary was forced to remove pots before leaving port.

"Sure. See you then," Gary said to Charlie in the hotel.

But the next morning when Charlie got down to the pier, the *Big Valley* was gone. Not long after, he heard the news: The ship had sunk. Gary was dead, along with four of his five deckhands.

The lone survivor was a crewman who had been asleep in his bunk when the *Big Valley* rolled over on its side—and failed to roll back up. The man had his survival suit with him in his cabin and put it on before he went up on deck. He got into the water and found his way to the boat's life raft. A few hours later, a Coast Guard HH-60 Jayhawk predeployed to St. Paul plucked him out of the ocean. Two bodies were recovered soon after: One fisherman had failed to get the hood of his survival suit over his head; the second never got his suit zipped up all the way. The other three men, including Gary, were never found. The survivor was the only one of the men on board who had taken a safety course.

It didn't take long to count the pots back at the dock. The *Big Valley*'s stability booklet approved the ninety-two-foot ship to carry thirty-one 600-pound crab pots on deck, and 5,000 pounds of bait below. The crabber left Dutch with fifty-five pots, each of them closer to 700 pounds. It was carrying almost

11,000 pounds of bait. What an asshole, Charlie thought. W[illegible] an idiot. Gary sure was a character, though. People would miss him. They'd say he ran into shit luck, and probably that God took him before his time.

THE MOON WAS ALMOST FULL over the cutter *Munro*. On the lookout deck, two crewmen scanned the modest waves. They'd grown bigger in recent hours, but it was still relatively calm near the ice edge. It was 2:52 A.M. when a petty officer standing night duty answered the phone in Combat, a cramped compartment in the belly of the ship lit mostly by the green glow of computer screens. It was District Command in Juneau: A large trawler with forty-seven people on board was taking on water approximately 150 miles south of the *Munro*'s position.

Less than a minute later, Operations Specialist Erin Lopez was shaken awake in her rack.

"OS1, we have a Mayday case!" the young seaman told her. "You need to come down." Like Ops Boss Jimmy Terrell, Lopez was from Texas. The twenty-six-year-old was the only person on the *Munro* who had graduated from the Coast Guard's Maritime Search and Rescue Planning School. Tonight she'd be the ship's Air Direction Controller (ADC)—in charge of the radios and communications with the Coast Guard aircraft. In seconds, Lopez was out of her bunk. Without changing out of her pajamas, she slipped into her boat shoes and ran the one level down to Combat.

Lopez's boss, Chief Luke Cutburth, called her a "SAR dog," a moniker she accepted as a badge of honor. It was obvious to Cutburth that the younger OS ate, slept, and breathed search and rescue. Lopez loved the strategy of juggling data to form a rescue plan. She thrived under pressure. Cutburth could relate.

In his off time, he competed in pro mountain-bike competitions. Meanwhile, Lopez was training for a half-Ironman. She is tiny, a self-described "five foot nothin'," with long brown hair that she pulled back in a firm ponytail. She'd secured Coast Guard sponsorship for her upcoming race and impressed some of the less coordinated crew members with her endurance on the *Munro*'s single treadmill. The gym was in the very bottom of the ship's bow. It was easy enough to get vertigo just standing down there, but Lopez managed to balance on the jolting exercise machine for hours at a time. Back in Kodiak, she taught a weekly spinning class at the air station's gym. She'd spent the previous evening helping to make Easter goody bags for the crew. The effort had been planned far in advance, with a shopping trip to Kodiak's Wal-Mart before the *Munro* left port. She'd done it the year before, too, pacing every level of the ship early on that Sunday morning to leave the homemade candy packets next to each pillow. It was a small thing, but she knew it'd make people happy to wake up on Easter morning to that little surprise.

Lopez opened the door to the dark control room. Formally, the space was known as the Combat Information Center (CIC) but most of the crew preferred just "Combat." The dark, cool cabin is the *Munro*'s war room, the brain of the ship. It is where they collect intel on nearby vessels to prepare for boardings, control the helicopter operations, and make decisions on any law enforcement or search and rescue mission. It was Operations Boss Jimmy Terrell's home—and Erin Lopez's.

The watchstander was on the radio with an officer from the sinking ship, the *Alaska Ranger*. Lopez took over. Could he tell anything about the rate of flooding? she asked.

It was in the rudder room. They couldn't keep up with it—the water was rising too fast. They'd already given up on the pumps and the fishing vessel's crew members were in their survival suits.

The voice on the other end of the transmission was clear and even, but Lopez knew that didn't mean the situation was under control. She'd worked for four years as a search and rescue specialist on the East Coast, and had already spent close to two years on the *Munro*. Most often when you got a panicked voice screaming "Mayday, Mayday!" it was some rube who'd run out of gas. The more experienced captains—and they were mostly experienced guys up here in Alaska—would be more likely to calmly report "Uh, Coast Guard, we got a little problem here"—even when their boat was already halfway underwater.

By 2:55 A.M., the *Munro*'s watchstanders had finished copying down the critical information and made the necessary calls. They called Jimmy Terrell, Captain Lloyd, and the engine room. Less than five minutes later, Lopez felt the *Munro* jolt and then bound forward at flank speed. The ship's engineers had switched the *Munro* from its standard Fairbanks Morse diesel engines to two Pratt and Whitney jet engines, huge turbines similar to those that power a 707 airplane. The turbines would suck down two thousand gallons of fuel an hour, as opposed to the two hundred gallons burned by the standard diesels. On the diesels the ship maxed out at 17 knots. The turbines could deliver 27—officially. With the wind and waves blowing their way and the engineers pushing their equipment for everything it had, the ship was soon speeding south at 30 knots, close to 35 miles per hour. They were four and a half hours away from the *Alaska Ranger*. As planned, the sprint would all be downwind.

LIEUTENANTS TJ SCHMITZ AND GREG GEDEMER were asleep in their racks when the phone rang in their tiny windowless cabin. The two men shared a four-man berth on the 02 deck of the ship, one level below the bridge and just aft, or behind, the captain's private stateroom. They had a private toilet and shower, a luxury on a ship. Both Schmitz, thirty-nine, and Gedemer, thirty-three, were helicopter pilots, trained to fly the Coast Guard's HH-65 Dolphin. (The first letter indicates the mission: in this case *H* is the military's code for search and rescue. The second *H* stands for "helicopter." The number represents that the Dolphin was the sixty-fifth helicopter design, or model, accepted by the U.S. military.) A smaller, lighter aircraft than the HH-60 Jayhawk, the Dolphin was slight enough to take off and land from the *Munro*'s basketball-court-size deck.

Schmitz and Gedemer were part of Kodiak's ALPAT shop, short for Alaska Patrol, and that's why they were stationed on the island. The larger Jayhawk helicopter was always the first aircraft to respond to search and rescue calls directly from Kodiak—it was bigger, tougher, and had a much greater range. In Alaska, range is probably the most important attribute of an effective search and rescue vehicle. Of course, if for some reason every 60 aircraft were already in use, a 65 team could launch on a call from Kodiak. For the most part, though, the 65 crews were in Kodiak to train and prepare for deployment on Coast Guard ships patrolling the North Pacific.

Every winter fishing season, two ALPAT pilots, along with a team of several ALPAT flight mechanics and one ALPAT rescue swimmer, were stationed on board Coast Guard cutters during their Bering Sea patrols. Few men looked forward to the winter assignment, though the *Munro* was better than most ships. Schmitz had told the scheduler that he'd rather spend sixty days on the *Munro* than forty-five on another boat. He got his wish. This deployment would be his last during his time in Alaska. He'd be partnered with Gedemer, who had arrived in the state just a few months before.

Schmitz had served for a decade as an Army pilot before joining the Coast Guard. He'd been stationed in Bosnia in the late 1990s while his wife and baby daughter were back in the States. When he got home, Schmitz put in an application with the Coast Guard. It wasn't an unusual move. More than a quarter of all Coast Guard aviators are "prior service," meaning they have previously served in the Army, Navy, or Air Force. The months at sea weren't easy on Schmitz's wife and kids, who were left alone in a small, cold Alaskan town. Still, it could have been much harder. It could have been eighteen months in Iraq.

SCHMITZ HAD BEEN UP LATE the night before. He was waiting for word that Shawn Tripp's Jayhawk crew was safely back in St. Paul. If the larger helicopter got into trouble between St. Paul and Dutch Harbor, the *Munro*'s 65 Dolphin would be the rescue aircraft.

Meanwhile, Gedemer had gone to bed early. Saturday was "morale night" on the *Munro,* which meant pizza for dinner, often creatively prepared by a group of volunteers from the crew. Then there'd often be some all-crew activity scheduled: a casino night or bingo game. There was no real gambling allowed on the ship—and no alcohol permitted on board. They'd have prizes, though. Tonight, the incentive was a good one: a free dinner out in Dutch Harbor.

Erin Lopez was the cochair of the morale committee. The evening's activity, though, wasn't her idea. It came directly from Captain Lloyd. He was a big supporter of the morale efforts and made some himself. The captain was known to throw on a cook's white shirt on Sunday mornings, stand behind the griddle in the galley, and take custom-omelet orders from the crew. After dinner on Saturday, the pipe came down that the crew should report to the mess deck for "The Number-10 Can Challenge."

The rules were simple. Three-person teams would volunteer to get a paint-bucket-size can of food from the ship's galley—with the label removed. When the captain said "go," the team would open the can and start eating. The first team with an empty can would win the contest.

One of the ALPAT mechanics, Logan Cole, wanted to organize a team. TJ Schmitz bowed out: No way was he doing it. He'd watch. Cole tried to convince another ALPAT flight mechanic, Al Musgrave. "No, no," Musgrave said. He didn't want to be the one to let the team down if they got something nasty. But pilot Greg Gedemer was game, as was Abram (Abe) Heller, the ALPAT crew's twenty-three-year-old rescue swimmer. The trio pried off the oversized can top to face a vat of cold baked beans. Gedemer must have wolfed down 3 pounds. It looked a lot better than what some of the other groups ended up with: pickles, potatoes, apple-pie filling, and—worst of all, Gedemer thought—beets. That team had to drink the juice as well. The ALPAT crew came in third out of eight teams.

Six hours after the contest, Gedemer was woken up by the ship's phone. He rolled out of his rack and picked up. It was Ops Boss Terrell. Gedemer handed the receiver to Schmitz, who as the more senior pilot was the aircraft commander on this patrol. Soon after, he felt the ship jolt forward and start tearing south through the swells with a high-pitched whistle. They were up on "the birds," the ship's 18,000-horsepower turbine engines.

Schmitz was the first one out the door and down to Combat to meet Terrell. Gedemer followed a few minutes later. "Do you know something I don't?" Schmitz asked the younger pilot when he showed up in the control center. Gedemer was already dressed out in his orange dry suit.

Schmitz stayed in Combat while Gedemer went to wake up the flight mechanics, who slept in a ten-man berth three flights down from the pilots' cabin. Then he made his way aft toward the hangar. The place was a mess. The ALPAT shop had their own small lounge just forward of the flight deck: It served as their workspace, their storage locker, and their clubhouse. They'd been watching DVDs of the television show *Heroes* in there the night before. The door of the small ALPAT fridge had busted open and food was everywhere. Cereal had spilled all over the floor. Gedemer had spent his first couple years out of boot camp on Coast Guard ships. You always think you're secured for sea, he remembered, until the first big storm. In this case, it was a man-made storm: the ship on turbines in rough seas. He started straightening things up, securing their snacks and extra gear.

Around the corner in the hangar, the four mechanics were beginning to prep the aircraft. Greg Beck was the head guy, then there was Logan Cole, Al Musgrave, and a newer mech who wasn't Alaska-qualified yet. He'd help out in the hangar, but he wouldn't be flying on any real cases on this patrol. The mechanics had come up with their own rotation schedule. One man was "on" until he flew, either in training or on a case. Then it was the next man's turn. They were free to arrange the system for themselves, and this seemed the fairest, especially in the Bering Sea, where they might easily go a whole week when the conditions were outside the limits for training flights.

Al Musgrave was happy with the system. In his experience, Coast Guard people were pretty good about coming up with a plan that made sense and was fair. Musgrave was from Barbourville, Kentucky. It wasn't the type of place where kids thought about joining the Coast Guard or where many people had even heard much about the service. Musgrave graduated from high school in 1997 and enrolled in the engineering program at the University of Louisville, three hours from his hometown. By the end of his freshman year, he felt directionless. He was partying more than studying. He paid a visit to the city's Coast Guard recruiting office. If the country had been at war, he probably would have dropped out of school and joined the Army or the Marines. But it was the late 1990s, and things were pretty quiet. The Coast Guard sounded good. Wartime, peacetime, it didn't matter: Coasties did their jobs every day. Six months later, Musgrave was at boot camp in Cape May.

His first assignment was a year-long detail on the *Midgett,* a 378-foot cutter based in Seattle. He timed it right: The ship was slated for the next over-the-horizon deployment, to the Persian Gulf to enforce the U.S. trade embargoes against Iraq. It was a pretty great job for a kid two years out of high school. The ship made port calls across the Middle East and Southeast Asia; the seamen were able to walk around foreign cities and try interesting foods. Musgrave hung around with the ship's ALPAT crew. By the time he got back to Seattle, he'd decided he wanted to be a helicopter flight mechanic.

Musgrave went to A School—the several-month-long training program that qualifies a new Coastie for a specific job—and afterward found himself stationed in North Bend, Oregon. There wasn't much action there. The small boat stations handled most of the rescues on the rugged Oregon coast. Musgrave's aircraft would often just be hovering above, taking video for the Coast Guard's public relations department. A couple times he helped deliver a pump to a boat taking on water. Once he rescued a surfer stranded on a rock after the ocean got a little too big for him. He'd never really been involved in anything major, though.

By the time Musgrave got his orders north to Kodiak in 2004, he was married with a little girl. The family moved into a house on base and joined the Mormon church in town. Musgrave was a woodworker and there was a shop he could use for bigger projects. He built bedroom furniture for his oldest daughter, and two more children who were born after he and his wife moved to Alaska.

Musgrave missed his family during his months at sea. On the ships, the ALPAT crew spent quite a few hours sitting around, waiting for something to happen. They watched a lot of movies. Most days, Musgrave spent an hour or two working out with a 200-pound sand bag in the hangar. He would sometimes go down and help wash dishes in the galley. You didn't need a special qualification to scrub pots, and it helped to pass the time.

BACK IN COMBAT, SCHMITZ AND GEDEMER had been listening in on the radio communications between the 60 Jayhawk helicopter, already airborne from St. Paul Island, and the officers on board the *Alaska Ranger*. Captain Lloyd was listening in, too, along with Terrell, Cutburth, and Lopez. The captain towered over the rest of them. He was six foot six, thin and fit, with gray hair that made him look distinguished and slightly older than his forty-three years. Craig Lloyd was a lifetime Coastie. Both of his parents had served in the Coast Guard. His wife was in the service, and so was his brother. He'd seen a lot: He had played a key organizational role during Hurricane Katrina back in 2005, and had led his share of cold-water search and rescue cases. He could already tell that tonight would be one he—and his 160-person crew—wouldn't soon forget.

Soon after 5:00 A.M., the *Munro*'s crew held a preflight brief. They reviewed the weather conditions, the information they had about the sinking ship, the geographic plan, and the objectives of the mission. Then, led by Erin Lopez, the crew drew up a GAR model. The acronym stood for both General Analysis of Risk and Green, Amber, Red, and it was an exercise the ship's crew completed before beginning any operational mission. Common sense in a bucket, Captain Lloyd called it. There were six categories to evaluate: planning, supervision, equipment, mission complexity, crew fatigue, and crew selection. The staff would assign each a number between one and ten. The higher the total number, the higher the risk: twenty-three and lower is considered green, or low-risk; forty-four and higher is considered red, or high-risk, and requires approval from District Command to pursue.

Lopez asked Schmitz to assess event complexity. The pilot thought about the distances involved, the long night (sunrise wasn't until 9:07 A.M.) and the sea state. They'd be dealing with high winds, snow, and—potentially, it sounded like—multiple victims in the water. Schmitz rated event complexity a ten. The GAR model revealed what was already obvious to Schmitz, Lopez, and just about everyone else in the room: They were about to take on a very high-risk, very high-reward mission.

The captain turned to Schmitz. Where did he feel comfortable launching?

"Eighty miles from the sinking site," the pilot said. It was farther than anything they would ever do in training, even in perfect weather. But he felt it was reasonable. The ship would be closing the distance, which meant the return trip should be shorter.

THREE DECKS UP FROM COMBAT, Musgrave and the other flight mechs were busy preparing the helicopter. When not in use, the forty-five-foot Dolphin was stored inside the U-shaped hangar at the rear of the ship. To fit inside the garagelike structure, the helicopter's rotor blades were folded in while the helo was still out on the flight deck, each along a hinge in the center of the blade. The bird was pulled back and forth from deck to hangar by a team of specially trained seamen known as the tie-down crew. Heavy canvas straps were secured to each corner of the aircraft and then cinched down inside the hangar or to holds on deck. No matter how rough the seas, the helicopter should be secure.

At the word from Combat, the tie-downs began to move the helo out onto the pitching deck. As the four-man team traversed the helo in the darkness, they could see the tips of the waves, whitecaps rushing by them at speeds they'd rarely—if ever—seen. Freezing spray pelted their backs and the backs of their heads, the cold water working its way toward any millimeter of exposed skin.

Slowly they tugged the 6,500-pound aircraft to the center of the platform, directly over a honeycombed metal platform known as the talon grid. A hydraulic arm, the talon, was lowered from the belly of the helo and latched on to the grid. The aircraft was safe.

The crew went back inside the hangar.

Now they'd wait.

Chapter Five

"Abandon Ship!"

Evan Holmes had his survival suit on and was ready to go. As a member of the emergency squad, the twenty-five-year-old factory manager was responsible for one of the life rafts and for helping the crew to abandon ship in an emergency. Of course, he'd never launched a raft for real, never done anything more than stand around the sealed canister and talk about what they would do in an abandon ship situation.

Together with a few other guys, Evan grabbed the ship's Jacob's Ladders, strong yet flexible ladders made of line and wooden dowels. They secured the ladders to the rail—one near each life raft—and hung them down over the side toward the churning seas. Once the rafts were launched, they'd climb down the ladders to reach them. But for right now, there was nothing to do but wait.

Eric Haynes was inside the wheelhouse. The glass on the wheelhouse windows was iced up, but the windows had a couple of small circles of thicker glass that remained clear even in the worst weather. Through one of the circles Eric could see Marco Carrillo smiling and waving to the people inside. Eric looked at the blurry line of red figures out on the deck. If they were really getting off the boat, he thought, they'd better not be hypothermic before that happened.

"Pete," he said to the captain, "how about if we do a rotation? Bring a few guys in at a time, just to warm them up. I'll tell them to keep quiet." Pete agreed, bring them in, but tell them to keep it down. No smoking. And tell the guys to turn their lights off, the captain added. Each of the ship's survival suits was equipped with a small strobe light. Many of the men had turned theirs on as soon as they got into the suits.

"They need to conserve the batteries," the captain said to Eric. "Tell them to turn those lights off!"

THE INDIVIDUAL STROBES WERE PART of a new Coast Guard program called the Alternative Compliance Safety Agreement (ACSA), designed specifically for the Bering Sea head-and-gut fleet. Like the dockside exams that had reduced casualties in the crab fleet since the late 1990s, ACSA had been designed to address safety problems within a very specific group of boats. And like the crab initiative, the ACSA program had been spearheaded by Coast Guard Commander Chris Woodley, with civilian Coastie Charlie Medlicott as his wingman.

There were just over sixty head-and-gut boats sailing out of Dutch Harbor. Like the FCA boats, most of them were owned by companies headquartered in Seattle (85 percent of all fish harvested in Alaska is caught by boats with owners in Washington State). The head-and-gut ships operated with the same unregulated status as little mom-and-pop catcher boats—even though the H&G fleet had proven in recent years to be much more hazardous.

Like with the crab boats, there were some obvious reasons why: The H&G vessels were big, most of them between 100 and 250 feet. All had treacherous processing equipment on board and enormous freezers that were cooled with dangerous chemicals like ammonia and Freon. On most boats, the frozen fish was packed into waxed cardboard cartons, packaging that had proven in the past to be a fire hazard.

The head-and-gut boats sailed with large crews of up to fifty people. Most of those men (and, with rare exceptions, they were all men) were working in the factory, not as full-time deckhands. It was common for the H&G boats to hire workers who were new to fishing and had little to no experience with boats or cold-weather hazards. The ships tended to sail with multicultural crews; cultural and language differences could cause problems during emergencies when quick, precise communication was critical. The ships regularly spent long periods at sea, weeks or even a month or more at a time. It followed that they were often far from port, far from shore—far from rescue should something go wrong. The combined hazards concerned some people in the Coast Guard, especially given recent casualties.

The ninety-two-foot H&G trawler *Arctic Rose* was home-ported in Seattle, and sailed out of Dutch Harbor. At 3:35 A.M. on April 2, 2001, the Coast Guard received a hit from the ship's EPIRB, transmitted from a spot several hundred miles northwest of St. Paul Island. The Coasties tried to hail her, with no response. A Hercules C-130 from Air Station Kodiak was launched around 4:00 A.M. and arrived at the EPIRB's coordinates at 8:40 A.M. The search plane identified an oil sheen and a large debris field. Another ship owned by the same company arrived around the same time and spotted a single person in the water, a man in a bright red immersion suit. A crew member jumped into the ocean to recover the man, and the body was hauled up on deck. The crew recognized him as the *Arctic Rose*'s captain, David Rundall. The fishermen attempted CPR on Rundall but with no success. The captain's survival suit was flooded with seawater.

Seven more suits were later found amid the debris—all empty. The *Arctic Rose*'s life raft was spotted floating upright and vacant. Rundall's body was the only one to be recovered. The boat wasn't fishing at the time of the disaster; most likely, the majority of the other fourteen men on board were asleep when the tragedy occurred. With fifteen fatalities, the *Arctic Rose* sinking was the deadliest single fishing vessel casualty in the United States in fifty years. The Coast Guard investigation into the sinking later found that of the fifteen men lost, nine were new processors who had been in the job for less than a year. Three of the dead were Mexican nationals working under assumed names.

Whatever happened to the *Arctic Rose* happened fast. There was no Mayday call. The debris provided few clues as to what had caused the tragedy, though a later examination of the sunken ship by an unmanned ROV (remotely operated vehicle) revealed that the aft weather-tight door that led from the stern deck to the processing space had been left open at the time of the sinking. What was known was that the tiny *Arctic Rose* had a reputation as one of the most poorly maintained boats in the H&G fleet (and the Coast Guard investigation found that the ship's owners had ignored many of the stability requirements mandated by its marine architects). Many in Dutch weren't surprised by her sinking. More than a few people wondered why someone in the Coast Guard hadn't stopped that "piss pot" from leaving port.

A year and a half later, there was another major casualty in the head-and-gut fleet, on the 180-foot freezer long-liner *Galaxy,* which had been built as a Navy minelayer in 1942 and converted to a fish-processing vessel in 1997. It started with a fire in the engine room, and a backdraft explosion that blew several crew overboard. While a rescue effort for those men was under way, a massive fireball from the engine room vents set the wheelhouse on fire and separated most of the rest of the twenty-six-person crew from their survival suits. Heroically, the captain remained in the burning wheelhouse long enough to get out a Mayday call and to help launch a life raft from the ship's top deck. He suffered severe burns in the process, but he lived—as did all but three of the other people on board.

There was some significant luck involved: Fifteen of the crew members managed to jump into the ship's just barely launched raft and were rescued by a Good Samaritan vessel about an hour and a half later. Three more people were rescued directly from the water by Good Sams, also within a couple hours of the sinking. Finally, Coast Guard rescuers were nearby, predeployed in Cold Bay, at the western tip of the Alaskan Peninsula, for the fall red king crab season. Five *Galaxy* crew members were successfully airlifted from the burning ship by a Jayhawk helicopter in the final hour before the sixty-year-old boat disappeared beneath the waves. Several of them had to leap into the water first to escape potentially toxic smoke and ongoing explosions.

It took several years for the Coast Guard to release their final reports on the sinking of the two ships. When they did, there was pressure to improve safety in the rest of the head-and-gut fleet. Many people in the Coast Guard thought the service didn't have the power to do much without new federal legislation; these were uninspected boats, after all.

Chris Woodley saw things differently. He found a loophole in the law, one that would allow the Coast Guard to improve safety in the H&G fleet, without waiting for Congress to act.

WOODLEY HAD BEEN THE LEAD INVESTIGATOR into the *Galaxy* sinking and was also involved in the *Arctic Rose* proceedings. Both investigations had found that the lost ships were doing more processing than they were regulated to do. Under U.S. Code, the H&G fleet had very specific products they were permitted to make. They were allowed to cut the head off a fish, but not the tail. They were allowed to remove a fish's entrails, but not to package and sell what were referred to as ancillary products, like fish eggs or roe. There was little logic to the regulations. There was no added danger in gathering the eggs from the refuse. Leaving them there was just money out the shit chute, as more than one fisherman had pointed out. And did cutting the tail off a fish rather than the head somehow justify more government regulation? Nope, it really didn't make much sense. Coasties and fishermen were in agreement about that.

But the law could be used to make everyone safer nonetheless, Woodley realized.

He started digging into the records. National Marine Fisheries Service data clearly showed that just about all the sixty-some-odd vessels in the Dutch Harbor head-and-gut fleet were selling products that only processing boats were technically allowed to make. In order to keep making those products—legally—they needed to be categorized as processing vessels, which meant they needed to be classed and load-lined. They were breaking the law, though it was a law no one had bothered to enforce.

At least not yet. Woodley saw the Coast Guard's in.

The vessel owners were invited to a series of meetings, and the Coast Guard explained their plan. The new program was voluntary, the owners were told. The choice was either stop making ancillary products and become a true head-and-gut-only vessel (what they'd said they were all along), or join the Alternative Compliance program. With the Coast Guard's help, their ships would get safer. And, with the Coast Guard's ACSA certification, they'd legally be allowed to keep making all their products without worrying about a future crackdown on enforcement.

It was a compromise—a pragmatic approach to a sticky problem. The Coasties understood that many of the H&G ships simply could not become classed and load-lined. Most of the independent societies that performed those tests would not accept boats over a certain age that had not been previously certified. Instead, the Coast Guard would institute its own safety program. The philosophy was separate but equal, or as close to equal as possible.

For the first time, Coast Guard fishing vessel examiners would be boarding the H&G boats at dry dock, examining their hulls and their watertight doors, and truly looking at their structural integrity. There'd be new, additional standards for training and for safety equipment—including the strobe lights on the survival suits.

Almost all the boat owners decided to join the program. Some needed to keep making the ancillary products to stay in the black, but, in Woodley's view, many of the owners were honestly attracted to a program that would make their ships safer.

Ship owners learned of the program by late 2005. They had to opt in or out by mid-2006. The Fishing Company of Alaska opted in. All seven of their boats—five trawlers and two longliners—would join the ACSA program.

In the couple years after, the expense of joining ACSA had been made plain. By spring 2008, an estimated $40 million had been spent fleetwide on upgrades mandated by the program. So many boats were participating that it was difficult for the Coast Guard to keep up. They'd even gone so far as to send examiners to a shipyard in Japan, where the FCA dry-docked their boats. There weren't any other companies that serviced their ships outside of the United States. To the inspectors, it seemed like a lot of time and expense to sail a ship all the way to Japan for dry dock work. Yet that was the way the FCA did it, and so the Coast Guard went along. In early winter 2007, the *Alaska Ranger,* as well as the *Warrior,* the *Juris,* and the *Spirit,* had spent several weeks at a Japanese shipyard. There'd been a long checklist of improvements that needed to be made. The ship sailed back to Dutch Harbor with about half of the work undone. The amount involved was more than the company—or the Coast Guard—had anticipated.

When the ACSA program officially got off the ground in 2006, it was agreed the final deadline for compliance would be January 1, 2008. By that date, all the H&G vessels enrolled should have reached an equivalent standard to the load-lined boats, according to Coast Guard documents. But several months before the deadline, it was clear to Coast Guard inspectors that the original date was way too ambitious. More boats had enrolled than originally anticipated, straining Coast Guard resources. The program was a collaboration between two Coast Guard districts: District 13, headquartered in Seattle, and District 17, which includes all of Alaska. Neither district had much spare manpower to devote to the new program. Across the board, the ships needed more work than the Coast Guard inspectors had expected. In some cases, it had been impossible for the ships to schedule dry dock time given the increased demand. There were really only a handful of shipyards in Alaska and Washington

State equipped to keep one-hundred-plus-foot boats in dry dock for weeks at a time. And of course, most boat owners wanted to schedule dry dock time during the same times of year, when the fishing grounds were closed.

By late 2007, the Coast Guard was informally letting ship owners know that there would most likely be another six months to get things in order. They were making progress, and that was what was important. Woodley felt that most of the boat owners were making a good faith effort. Certainly, most of the ships in the fleet were in better shape than they'd been just a couple years before—and much safer than they would have been if the Coasties had just let things stand as they were.

Julio Morales was on the narrow deck in front of the wheelhouse, holding tight to the ship's rail. He'd been scanning the ship for his cousins, Byron and Marco. Byron was assigned to a bunk room one deck lower than the one Marco and Julio shared with six other men. He hadn't seen his cousin when he was in the wheelhouse, getting on his survival suit. But he spotted him now, about fifteen feet away, with a group of men leaning against the wheelhouse windows on the ship's port side. He recognized him because of his hair. Byron had the hood up on his survival suit, but his shoulder-length hair was all over his face.

"Byron!" Julio yelled across the deck. "Put your hair in. Under the hood!"

Julio motioned along the seam of his own survival suit. If Byron went in the water, he wouldn't be able to see.

"Tuck it in!" Julio yelled again.

Byron wasn't listening. He looked scared. Julio was scared, too. They'd been shown how to put on the strange suits, but that was it. Now what were they supposed to do? They didn't know how to get off the boat or what to do once they were in the water. He didn't want to let go of the boat to go help Byron. It wasn't safe; the deck was too icy. He'd been told he should stay with his muster group on the starboard side of the upper deck.

BACK INSIDE THE WHEELHOUSE, David Silveira was still locked on the radio, updating the Coast Guard and other FCA ships in the area, including the trawlers *Spirit* and *Warrior*. The *Warrior* was the closest—less than four hours away.

At 3:06 A.M., Kodiak watchstander David Seidl radioed the foundering vessel again.

"*Alaska Ranger,* this is COMMSTA. Request you confirm that you have set off your EPIRB, over."

"Roger," Silveira answered, the EPIRB was transmitting. "We lost a rudder," the first mate told Seidl. "That's where the water was coming in."

"*Alaska Ranger,* this is COMMSTA," Seidl replied. "Understand you have set your EPIRB off, you lost your rudder, that's where the flooding is coming in. Also be advised, the Coast Guard cutter *Munro* is en route with an ETA unknown at this time, over."

"Roger, roger," Silveira answered.

COAST GUARD LIEUTENANT TOMMY WALLIN was asleep in military billeting at Elmendorf Air Force Base in Anchorage when the phone rang, at exactly 3:00 A.M. The operations center in Kodiak reported that a fishing boat was in trouble 140 miles west of Dutch Harbor. A 60 Jayhawk crew predeployed to St. Paul Island would be launching soon. The C-130 crew was needed to provide backup for that aircraft.

Wallin wrote down the position.

First introduced in the mid-1950s, the ninety-seven-foot-long Lockheed C-130 is a versatile aircraft that's popular among militaries worldwide; the four-engine turboprop has been used as a gunship, for troop and supply transport, medical evacuation, and aerial firefighting. In 2008 the Coast Guard had thirty-three C-130s in service, five of them in Kodiak. The back of the plane is designed to adapt to the needs of the moment. Commercial-airplane-style seats can be installed to transport staff or the occasional television crew up for an Arctic awareness trip to check out the decaying state of the sea ice and report on the Coast Guard's plans to expand into the ever-growing northern seas. A removable computer bank with high-powered cameras makes it possible for technicians on the plane to study vessels or debris in the water below on TV screens on board. It comes in handy for fisheries patrols, when the air crew are often trying to read the name on a ship's stern, or to identify the gear on deck.

The rear of the plane opens up in a ramp to the ground, allowing the aircraft to fly a full-size SUV out to St. Paul, Attu, Cold Bay, or any of the more remote stations. Even the 65 Dolphin helicopter can be disassembled and transported in the Herc's bay, though reassembling the little bird is a multiday chore that many mechanics dread. During search and rescue cases, the rear opening also allows the plane to drop rescue supplies to vessels—and people—in distress. The crew can "punch" a dewatering pump, a life raft, medical supplies, or a data marker buoy, an arrow-shaped orange float that broadcasts GPS coordinates and allows a rescue team to relocate an emergency site and to track the flow of debris with winds and currents.

The Herc is the Coast Guard's workhorse for the "search" part of search and rescue. But despite all the plane's capabilities, it can't actually lift anyone out of the water. That's a job for the helicopters. When a casualty site is known, the C-130's most important role is often flying cover for a helo crew. Usually the helicopter will be doing the work of dropping rafts, pumps, and supplies, and lifting any victims from a ship or from the ocean. The C-130 is there in case the helicopter gets into trouble. If the helo goes down, the plane will pinpoint the location and drop a raft.

It took just minutes for the seven-man C-130 crew to pack their bags and load into the van for the half-mile ride back to the aircraft. The plane's engines were still warm from the flight from Kodiak, and the aircraft was already fueled. But with the rescue site almost a thousand miles away, the crew decided to add another several thousand pounds of fuel.

The Herc crew knew the Jayhawk helicopter should reach the scene close to an hour before they did. It sounded like a fairly standard case. Most likely the helo crew would be dropping the ship a pump. That was usually how this type of thing would go.

ERIC HAYNES COULD HEAR THE *ALASKA RANGER*'S engines struggling. He was moving in and out of the wheelhouse, trying to help make room for more men to rotate through. "What's going on?" "Are we going to be all right?" the crew was asking Eric.

"The Coast Guard knows our location," Eric said, trying to reassure them.

Based on what he'd overheard, Eric said, it sounded like they'd be able to hold out. "The Coast Guard is on the way," he told his crewmates. "And the *Alaska Warrior* is coming, too."

The engines sounded like they were fully underwater. Someone said they'd lost steering. But there was still power. Eric could hear the ship's officers talking about whether to shut down the engines. It seemed like Captain Pete was against it. The captain was consulting mostly with the ship's assistant engineer, Rodney Lundy. Rodney had been on the *Ranger* for more than a decade. The other two engineers, including Chief Dan Cook, were in their first season on the ship. Dan was still advocating for an immediate abandon ship. Eric got the impression that everyone was listening to Rodney instead.

At 3:11 A.M., watchstander David Seidl radioed the ship once again. "*Alaska Ranger,* this is COMMSTA. Request to know how much fuel and what type of fuel you have aboard, over."

In any marine casualty, the Coast Guard works with state and federal environmental authorities to document and monitor any environmental damage. Already, the Coasties were thinking ahead to the worst-case scenario.

"We have . . . roughly one hundred forty-five thousand gallons. . . ." Silveira responded.

"*Alaska Ranger,* this is COMMSTA. Confirm one hundred forty-five thousand pounds of diesel, over."

"One hundred forty-five thousand *gallons,* okay?" the first mate clarified.

"*Alaska Ranger,* this is COMMSTA. Understand. Nothing further. Talk to you in five mikes, over." (The substitution of "mikes" for "minutes" is common in radio communication.)

A few minutes later, the ship's lights began to flutter. "We're going to lose them," Eric heard one of the ship's officers say. He stepped back out onto deck. The stern looked like it was completely underwater.

"We're about to lose power," Eric yelled to the men clustered around the rail. "The lights are going to go out. Don't panic!"

It was 3:23 A.M. when Silveira relayed the outage to COMMSTA Kodiak.

"COMMSTA Kodiak, *Alaska Ranger.*"

"*Alaska Ranger,* this is COMMSTA Kodiak, over."

"Yeah, COMMSTA Kodiak, *Alaska Ranger.* We just lost all the lights."

"*Alaska Ranger,* this is COMMSTA. Understand you have lost all your lights at this time, over."

"Roger that," Silveira answered. "We just got the emergency lights on right now, whatever flashlights we have."

A few minutes later, the *Alaska Ranger* got an ETA to their position from the *Warrior:* 6:30 A.M.

OUTSIDE, THE DECK WAS SLICK WITH ICE, and waves were beginning to crest over the stern. It was spitting snow and blowing hard. The temperature was no more than 15°F and the black water that waited for the men a couple of stories down was guaranteed to feel much, much colder.

Processor David Hull was leaning against the wheelhouse window, the bow of the ship directly in front of him, when the *Alaska Ranger* went dark.

Oddly, the boat seemed to shift into reverse.

There was a stench of diesel smoke coming from the stern, where the Atka fishing gear was piled in massive mounds. Several men watched as a formidable wave crashed over the rear of the ship and retreated with the *Ranger*'s trawl net in its grasp.

"There goes our million-dollar net!" someone yelled as the huge woven mass spread like a puddle on the surface of the water, and started drifting up the ship's starboard side.

Within seconds, the thirty-five-year-old trawler took a sudden, violent list to starboard. David felt the ground drop out from under him. He lunged for a rail and held tight as crew members clinging to the metal beneath him gazed up in horror. Twenty-two-year-old steward Jeremy Freitag was right below David.

"Don't let go, David! Don't let go!" Jeremy yelled as two men slid straight down the narrow deck, through the open rails—and into the ocean. If David lost his grip, he would hurtle down the rail like a bowling ball, knocking Jeremy and half a dozen more men right off the edge.

Jeremy ground his feet into the metal floor and locked his arms to the rail.

"Hold on!" he yelled again at David.

It was pitch-black, the wind whipping across the exposed deck. After a few more seconds the boat seemed to shift upright a bit. Still, the list was at least 30 degrees. People were yelling, "Man in the water! Man in the water!"

Jesus, those guys went straight through the rail, Jeremy thought.

Everyone was talking at the same time. There was a plume of diesel smoke wafting forward from the stern deck. Jeremy heard someone yell, "Abandon ship." He started thinking about a TV program he'd seen, about how people could get sucked down with a ship. If you were right next to the boat—or on deck—when it sank, the force might pull you under, too. *I've got to get away from the boat*, Jeremy thought. *I need to get far away, as fast as possible.*

David had his arms wrapped around the metal rail. He still had his computer bag slung around his body. Around him were several newer guys, among them thirty-one-year-old processor Alex Olivarez. David and Alex had been working together in the freezer all winter. Both men were from Washington State. A couple years before, Alex's little brother had been murdered. It was a gang killing, still unsolved.

After his brother's death, Alex had become deeply depressed. He was fired from his mill job. His mother was suffering, distraught over the loss of her son. For a couple years, Alex had been watching the reality show *Deadliest Catch*. Fishing would be a good way to make some fast cash, he thought. He could help his mother and maybe hire an investigator to find his brother's killer.

"I don't know how to swim!" Alex was yelling across the deck.

David saw the newer processor clinging to the rail. He looked terrified. The ship was still listed hard to starboard and draped in darkness.

"Alex, let's pray together," David said.

He knew Alex was religious.

"Yeah," Alex said. He bowed his head. David and several other men nearby did the same. "Will you help us with this tragedy?" Alex said aloud. "We're scared. We know some of us might die. Will you help us, God? Help us, and let the majority live."

THE SUDDEN LIST HAD ALSO STARTLED everyone inside the wheelhouse. Evan Holmes saw Captain Pete fall down against the carpeted floor as the ship took the sharp fall to starboard. The factory manager pulled him up, then helped the captain zip up the survival suit that was down around his waist.

"All right, Captain, we're going now?" he asked Pete.

"Yeah. We're going," the captain answered.

Evan raced out to the starboard side of the wheelhouse deck to his assigned life raft, number three. Eric Haynes was right behind him. The raft was stored inside a white, barrel-like container mounted right up against the rail on the ship's deck. The *Ranger*'s bow was elevated high above the water, as though a huge weight was pressing down against the stern trawl deck. Tiny balls of icy snow stung Evan's cheeks.

It felt like the ship could capsize at any moment.

First, he needed to tie the 110-foot-long painter line that was attached to the raft to the ship's rail, above where he'd earlier tied the Jacob's Ladder. The list was so bad that Evan felt like he was being pushed up against the metal bars.

The full moon was breaking in and out of the clouds, but Evan could barely see anything.

Meanwhile, Eric Haynes felt along the metal strap holding the raft's container shut. Eric couldn't see anything either; his eyes were still adjusting from the sudden loss of the *Ranger*'s bright lights. Years ago, they had regularly reviewed how to launch the rafts, but now as Eric squinted at the ice-encrusted raft nestled in its cradle, he was drawing a blank.

Finally, the ship's cook remembered the light on his suit, and turned it on. He leaned over the barrel-shaped case and there it was: the clip. You were supposed to hold down the clip then pull off a ring to release the life raft, Eric remembered. With all the ice, it took Eric both hands to push the clip in far enough. Another crewman popped off the ring, and the barrel cracked open, emitting the rubber life raft.

The raft inflated with the tug of the painter line, and swung out and down into the black water.

The plan had always been to launch the raft and then climb down the Jacob's Ladder from the side of the ship to get in. By pulling on the painter line, a crewman still on the vessel should have been able to hold the raft in place until everyone was inside. Then, the line would be cut.

But when Eric's raft hit the waves, it immediately shot forward, toward the *Ranger*'s bow.

"It's going too fast! It's going too fast!" someone yelled.

Eric stared. The raft was still in sight, but it was nowhere near the Jacob's Ladder.

Raft number one was also on the starboard side, about twenty feet closer to the bow. Evan watched as another group of men struggled to launch that raft. They, too, were having a hard time. Finally, he saw the forward raft inflate and swing out toward the water. Again, the raft bolted forward, toward the bow of the ship. Then it seemed to disappear.

Holy crap, Evan thought. One of the rafts is gone.

He started doing the math in his head. There were forty-seven on board. With one raft gone, they would just have to try to crowd everyone in the other two. Evan had tied a dozen granny knots in the painter line holding raft number three to the rail. "If you can't tie a knot, tie a lot," seasoned mariners sometimes joked about the knot-tying skills of newbie crew. In this case Evan had taken the saying seriously. This thing isn't going anywhere, he'd told himself. But now the raft was so far from the ladder. There were only a few crew members who had actually ever tried getting into a raft straight from the water. Evan had done it during his training in Seattle, but it'd been damn hard. And that was in daylight, in calm water that was a whole lot warmer than the Bering Sea.

Evan and Eric and a couple other crewmen grabbed onto the painter line. They pulled the rear starboard raft with all the strength they had, but it barely budged. It was as if they were on the stern platform of a water-ski boat, trying to pull a guy on a tube in toward the boat as it was skipping at high speed across the water. There was just no way, it wasn't happening.

The boat's list seemed to be increasing—or at least the ship seemed to be getting lower in the waves. Water was all the way up past the base of the gantry. Eric felt like the ship might flip over at any moment. He and Evan told the men at their muster station to start going down the Jacob's Ladder and to try to swim to the raft.

"Let's get going, guys! You're going in one way or another!" Eric yelled. "Keep going," he instructed as one man after another gripped the ladder and started down the side of the ship.

The men were quiet; they looked calm.

They were scared, Eric knew. He watched each man follow the ladder down and drop into the waves.

In seconds, each one was gone.

BACK ON THE FRONT SIDE OF THE WHEELHOUSE, Julio Morales was still hugging the rail. He'd been holding tight when the ship listed. He'd seen those men fall off and heard everybody yelling "Man overboard!" It seemed like everything had gone from calm to chaos in the moment that the ship tilted to its side. Julio had been thinking that the *Warrior* would get there in time to save everybody. Now, looking out at the empty horizon, he knew no one was coming. They were going into the water.

Other people were moving, but Julio just stayed put. He was thinking. The Coast Guard was on their way. He had overheard the officers making a Mayday call, and made out the words "U.S. Coast Guard" in the answer that came back. It would be better to wait until the last minute to get off the ship, he thought.

Julio could see that the men had launched the life rafts. They were far away from the side of the ship, but they were still in sight. Carefully, Julio made his way to the starboard rail. Byron was already there. There was a raft attached to the ship with a line, and

Julio saw a couple of guys grab on to the rope and follow it into the water. It seemed like they knew what they were doing.

"Grab the rope," Julio told Byron. "Grab it, follow it down!"

Julio watched from the tilted deck as his cousin grabbed onto the painter line. Byron had his feet against the outside of the ship, and was leaning back with the taut rope in his hands, like a rock climber rappelling down a sheer cliff face.

"*Ayúdame,* Julio!" Byron yelled back up to the deck. "Help me!"

He was only a few feet from the water. The list was so great that the distance from the upper deck to the surface of the ocean was no more than a single story. But Byron seemed stuck.

"Help me!" he yelled again in Spanish.

"How can I help you? Just follow the rope!" Julio yelled back. It was dark, and with the list, it was hard to see from the side of the ship into the waves. Already, it was only blinking strobe lights that allowed the men on deck to identify people in the water. Julio could see that more and more people were going in. Maybe three-quarters of the crew had already abandoned ship. He watched as several men jumped from farther back, on the starboard side.

When he looked down again, Byron was gone.

INSIDE THE WHEELHOUSE, DAVID SILVEIRA and Captain Pete Jacobsen were taking turns working the radios, talking to the Coast Guard and the other FCA ships. They knew several boats were steaming toward them with everything they had.

It was around 4:15 A.M.—an hour and a half after the original Mayday call—when the *Alaska Ranger* initiated a call to the Coast Guard.

"COMMSTA Kodiak, *Alaska Ranger.*"

"*Alaska Ranger, Alaska Ranger,* this is COMMSTA Kodiak, over," watchstander David Seidl answered.

"We are abandoning ship." It sounded like a different voice from the one Seidl had heard before.

"We are abandoning ship," the *Ranger*'s officer repeated. His voice was strained but calm.

"*Alaska Ranger,* this is COMMSTA Kodiak, roger. Confirm you are abandoning ship at this time, over."

"Roger. Roger."

"*Alaska Ranger,* this is COMMSTA. Understand you are abandoning ship. Request you keep your EPIRB with you, keep your EPIRB with you, over."

"Roger. Roger."

"*Alaska Ranger,* this is COMMSTA Kodiak. Also, be advised a rescue C-130 is airborne and en route to you guys at this time, over."

Again: "Roger. Roger."

"*Alaska Ranger,* this is COMMSTA," Seidl said. "Roger. Be safe. We'll be there when we can, over."

"That's a roger," came the reply, the voice weak across the eight-hundred-mile swath of sea.

Chapter Six

The Observers

Gwen Rains had been on board the *Alaska Ranger* for four days. For the past two years, she had worked on and off as a federal fisheries observer. In a given year, approximately 350 observers sail on fishing vessels in Alaskan waters, recording information about the catch. Their data are used to manage the fisheries in real time and to set annual quotas for different species.

Smaller vessels under 60 feet in length are exempt from the observer requirement. (Unsurprisingly, 59 feet has become a popular boat length and owners have been known to saw off a couple feet of bow to squeeze in under the limit.) Ships from 60 to 125 feet in length sail with an observer 30 percent of the time. Most large ships, including the *Alaska Ranger* and the rest of the FCA trawlers, sail with two observers every time they leave port.

The cost of running the Alaskan observer program is shouldered mostly by the fishing companies themselves. The price tag is more than $350 per day per observer, just less than half of which goes toward administrative costs. The federal government pays for the observers' training and for data management through the National Marine Fisheries Service (NMFS), which is an agency within the National Oceanic and Atmospheric Administration (NOAA).

The job isn't glamorous. Gwen learned that pretty quickly. An observer's main task at sea is to sample the ship's catch. The data are used to assess the overall health of the fishery, and to determine if the ship is observing environmental laws—among them reporting its catch accurately. The end goal is to keep Alaska's fisheries sustainable. Over a period of three months, an individual observer might spend time on one, two, three, or even four different boats.

Though the fisheries observer program exists across the United States and in many other countries, Alaska's system is more complex than most. There are a huge number of fisheries, each with its own set of complex regulations. The types of boats vary widely as well. An observer has to learn how to work effectively on long-liners, pot boats, and trawlers, on small boats with just four or five crew, and on huge processors with more than a hundred workers on board. Each ship may present a new gear type, a new daily schedule, and a new method of catching and—often—processing fish.

On each boat, Gwen had to figure out the best method for taking a scientifically sound, random sample of the catch. How much fish was the ship hauling up? What was the distribution of species? What prohibited species were being caught, and in what quantities? If time permitted, she would sex her fish samples by examining the gonads and determine age by extracting the otoliths, tiny ear bones whose annual ridges can be counted like tree rings. The work was messy and smelly. It took place either on the cold, wet deck of a catcher boat bouncing in the Alaskan waves or, on the larger processing boats, belowdecks in a frigid, damp factory reeking of fish—and fishermen.

The hours were long. But on a large vessel with two observers on board, like the *Ranger,* the shifts were relatively stable. Each observer worked twelve hours on, twelve hours off while the boat was actively fishing. (When the ship was transiting to or from the fishing grounds, there'd often be paperwork to do, or sleep to catch up on.) Each time the catch was hauled aboard, an observer had to be there to take a sample. On most boats, Gwen sent daily forms via e-mail from the ship to the program administrators at NMFS. Observer data can shut down a fishery if, for instance, sampling reveals that a large amount of prohibited species are being pulled up in the haul. In addition to keeping track of what the boat wants to catch (the "targeted" species) observers are responsible for documenting "prohibs" like halibut, crab, and salmon, whose harvest is strictly allocated to specific types of boats. A factory trawler is never allowed to keep these species; rather, they have to be thrown back into the ocean, alive or dead. They're usually dead.

Fisheries observers sometimes have to play the role of cops on ships, and that means they're occasionally treated with contempt. It wasn't an easy job, but Gwen loved it. Like most observers, she worked for a few months and then took a few months off. She noticed that during the time away, the drudgery of the job tended to fade from her memory, while the joys—the crisp summer days in Dutch Harbor, the late-night chats with captains in the rolling wheelhouse—stayed with her.

Gwen worked for Saltwater Inc., one of five private companies that contracted with NMFS to provide observers for the Alaskan fisheries. The starting pay was $130 a day. By 2008, Gwen had worked her way up to a $190 daily wage. Except for the odd meal out in Dutch Harbor, food and housing were paid for. She got an allowance for clothing, and the company paid her airfare to and from the fishing ports and back to Seattle, where she debriefed with a NMFS staff member before heading home. A lot of fishermen—and observers, too—blew a great deal of cash on alcohol when they were in Dutch, but Gwen wasn't a big drinker. She saved pretty much all she earned, about $15,000 in a three-month stint.

At thirty-eight, Gwen was older than the average fisheries observer. The prerequisites for the job include a four-year science degree, preferably in biology or marine biology, and at least one class in both math and statistics. Of approximately two hundred new observers trained in a given year, only half will come back after their first three-month contract. Of those remaining hundred, perhaps fifty will still be in the job a year later. Like many fishermen, most new observers arrive in their first Alaskan port having never spent significant time on a boat. The majority are in their early twenties and looking to make some money and get some real-world experience before applying to grad school.

Gwen didn't fit the obvious mold. She was from Marshall, Arkansas, a divorced mom with four kids at home—two boys and two girls, ranging in age from ten to seventeen. She'd dreamed of being a marine biologist since she was a child and had spent years working her way to a four-year biology degree from the University of Central Arkansas. In June 2006, she spotted the fisheries observer position on the job search Web site Monster.com. In July Gwen was on a plane to Anchorage. She'd struggled with the decision to leave her kids with their dad for the months she'd be in Alaska. But the job felt like the fulfillment of a lifelong dream. She decided to go.

In the two years since then, Gwen had worked on a variety of boats, including a couple owned by the Fishing Company of Alaska. She had probably spent half her total sea days on FCA boats. All of them had Japanese fish masters. Gwen had heard from other observers about Dutch Harbor boats with Norwegian fish masters. If a company hired a fish master, he was pretty much the number one person on the boat—the guy who'd be making the key calls. As Gwen saw it, the captain was essentially a taxicab driver.

Gwen was told to report to the *Ranger* in the third week of March 2008. She was waiting at the pier when the boat pulled up, at close to 2:00 in the morning. As Gwen boarded the ship for the first time, her friend Christina Craemer, a Saltwater observer who had been on the ship since A season began in mid-January, was getting off. Chris and Gwen had gone through the three-week observer training class together in Anchorage a year and half before, and they'd hit it off right away. They were both a bit older than the average student and were both from small farming communities. Gwen and Chris recognized an appealing, down-to-earth quality in each other. They were both runners, and enjoyed jogging together when they found themselves with the same off time in Dutch Harbor.

Chris told Gwen about the recent changes on the ship: how Captain Steve Slotvig had left the boat after a series of fights with the fish master and been replaced by Pete Jacobsen. Another FCA captain, David Silveira, had come on as first mate.

Gwen was happy with the news. Silveira normally sailed as the captain of the FCA long-liner *Alaska Pioneer,* and Gwen had been aboard that ship the previous two B seasons. She'd spent dozens of hours in the wheelhouse talking with the handsome former tuna fisherman and counted him among her favorite people in Dutch Harbor. Silveira was extremely charismatic.

He could be stern when he needed to be, but it was obvious he had a big heart. He was the type who often went out of his way to assist crew members with their personal problems—and to be helpful to observers.

It was part of Gwen's job to record any marine mammal sightings when at sea, but mostly she just loved to stare at the animals for as long as possible. Every time Silveira spotted a pod of whales from the *Pioneer*'s wheelhouse, he'd announce Gwen's name over the ship's loudspeaker. Silveira didn't want to distract his crew—many of whom would be handling dangerous hooks and lines—by announcing a whale sighting. But Gwen knew that when she heard Silveira say her name there was something to see.

Both Chris and Gwen knew that often the best way to get to know men on the ships was to ask them questions about their families. A lot of fishermen put on a tough guy persona. And there was plenty of truth to it: Many of the men working on the *Ranger* had criminal records. One quip around Dutch was that FCA was really an acronym for "Felons Cruising Alaska." Many of the fishermen would talk openly about the time they'd served, or about their history with drug or alcohol abuse. But once the observers got to know the individual guys, they usually found that most of them were hardworking men doing their best to make a living at something respectable. There was no denying that it tended to be a rough crowd on the bigger factory boats. There was going to be a lot of attitude, a lot of foul language, maybe a few fights. As an observer, you had to expect that. You were entering their world, after all.

Captain Pete, though, was a different type. In his mid-sixties, he was older than the rest of the crew. He was small and thin, with a gray beard that he kept neatly trimmed. His manner was calm—soft-spoken, even. He rarely yelled, which was not something that could be said of many captains. Pete Jacobsen was a kind man, and generous with his time. Earlier that winter, he had personally gone out and bought new carpet for the observers' room after noticing that it was particularly musty. The act was unprompted; no one had complained about it. The captain laid the new carpet down himself.

Pete was a neat freak and would personally sweep the wheelhouse every single day. Sometimes he would vacuum, too, even though he could have had someone else do it; it was someone else's job. Even on the ship, he dressed well, often wearing a button-up shirt with a pointed collar and snap buttons that his third wife, Patty, had made for him. At home in Lynnwood, Washington, Pete was the kind of husband who would spend all day working on a clogged drain, or searching the whole house to help Patty find a missing sock. After twenty years of marriage, Pete was still awed by the way his wife held their lives together while he was gone. It could be hard. He'd worked for the FCA since the start, and even though he told Patty it was the Japanese who really ran the company, he felt an intense loyalty to FCA owner Karena Adler. When the company called, Pete never said no. They might have plans for a trip the next day, but still he'd be headed back up to Dutch. Patty would be at home in Lynnwood, taking care of the kids and later the grandkids. Pete had two children with Patty, a stepson named Scott, and a daughter named Erica. And he had two children from his first marriage, Carl and Karen.

Pete hadn't been in good touch with his older children for many years. But more recently, he and his daughter Karen had grown closer. Pete was proud of the things she'd accomplished—her master's degree and her job as a nutritionist. Karen had become a devout Christian, and on the few occasions when she visited her father in Seattle, they went together to a Baptist church near his home. For her thirtieth birthday, Pete bought Karen a mariner's cross, a necklace with a pendant of Jesus on a ship's wheel and anchor. A few years later, Karen selected a leather-bound Bible for her father, and had "Captain Eric Peter Jacobsen" embossed on the soft cover.

Pete Jacobsen loved his family, that was clear. But for more than twenty years, he'd spent nine, ten—sometimes eleven—months out of the year in Alaska. Sometimes when Pete talked about his children, it seemed like he was talking about people he didn't know all that well.

LED BY CAPTAIN PETE, GWEN WALKED through the *Ranger* looking at safety gear. Each time she got under way on a new boat, Gwen was required to check that the ship had a current Coast Guard safety decal—a sticker issued to a vessel after a successful dockside exam—and to fill out a standardized safety checklist in her logbook. The efforts were for her own benefit: to determine that the ship was safe enough for her to board.

She checked out the ship's survival suits and the EPIRB. She noted that a number of the *Ranger*'s fire extinguishers did not appear to be in "good and serviceable condition" as her list stated they should be. Gwen's form was divided between "go" and "no-go" items. The Coast Guard decal was a no-go: If the boat didn't have one, or if it wasn't up-to-date, she couldn't sail on the ship.

That observer program rule had turned the Coast Guard's so-called voluntary dockside exam into a mandatory one, at least for ships over sixty feet. It made sense that NMFS wanted to protect its own people by ensuring that the boats they'd be working on had the proper safety equipment. But the decal rule begged the question: If a boat isn't safe enough for a government observer, why is it safe enough for a few dozen fisheries workers who are likely to have even *less* safety training? It was a double standard, one that seemed to say that the life of an observer was more worth protecting than the life of a fisherman.

The rule was a sore spot among those in the Coast Guard who had been arguing for mandatory inspections for years. It was embarrassing that the Coast Guard couldn't enforce inspections but that the fishery management body could—and in a very short amount of time, without public comment.

Gwen examined the decal and then the life rafts: They had to hold everyone on board and have up-to-date inspection stickers. Many of the other items on her list were discretionary. Technically, an observer has a right to refuse to get on any boat, though it rarely happens. If an observer rejects a boat for anything other than no-go reasons, the company has to scramble to find a replacement. Meanwhile, the ship can't leave port, which is a situation that's bound to make many people extremely unhappy.

As she walked the ship with Captain Pete, Gwen noticed that some of the seals on the *Ranger*'s watertight doors looked like they were in poor repair. A couple were frayed, and it was a struggle to get the doors closed correctly. The ship was dirtier than any other boat she'd been on. Still, the *Ranger* passed her checklist. As the ship got under way, she unpacked her bag in the cabin she'd share with her co-observer, Jayson Vallee.

JAY WAS TWENTY-FIVE YEARS OLD and newer to the job. He wore round, wire-rimmed glasses and had a full beard and coarse red hair he tamped down with a Red Sox baseball cap. Three months earlier a friend had dropped Jay off at the Manchester, New Hampshire, airport for the flight to Anchorage. It was Christmas Eve, and Jay had managed to get a one-way ticket from New Hampshire to Alaska for just $400. He'd have multiple stopovers, but the price was worth it. Saltwater would reimburse him $350, their estimate for the trip from Seattle to Alaska.

Jay had held a series of odd jobs since graduating with a four-year biology degree from the University of New Hampshire in 2005. He drove a bus and worked processing insurance claims. He saw Saltwater's ad on Craigslist, interviewed over the phone, and was offered the position. It wasn't a hard decision. He'd been out of college for almost two years and this was the first job offer he'd had that had anything to do with his major. He was on the flight a couple weeks later.

It was after 1:00 A.M. when Jay stepped off the plane into Anchorage's newly renamed Ted Stevens International Airport. He was expecting deep snow, but there'd been more accumulation in New Hampshire. He got in a cab. The driver already knew the way to the Saltwater bunkhouse.

In the early 1990s, when the federal observer program was just getting off the ground, Saltwater was among ten companies that formed to screen, hire, and manage new Alaskan observers. Since then, the number of contracting companies has been whittled down to five. Two of them, including Saltwater, have their headquarters in Anchorage. The other three are in Seattle. New observers can get their three-week orientation training in either city.

There were nineteen in Jay's class, eight men and eleven women. Every weekday for three weeks, they reported to the second floor of an unadorned office building in downtown Anchorage. The classroom where the students met had windows facing south. They got there by 8:00 A.M. It was after 10:00 by the time the sun rose behind the Chugach Range to the east of town, and long before they left the building at 5:00 P.M. each day, all traces of light had already disappeared below the horizon. It was depressing, Jay thought. It felt like the whole city was draped in a permanent gloom.

The space wasn't unlike the college science classrooms where most of the students had spent untold hours as undergraduates. Stuffed fish were hung on the wall. Crowded bookshelves held stained manuals and species-identification books. At the back of the room, the contents of a life raft's survival kit had been broken open and arranged on a corkboard wall. There were flashlights, flares, and first-aid supplies. Water and food rations were displayed, along with a paddle, a rigid bag used for bailing out a raft in rough seas, a large rewarming sack, and a horseshoe-size plastic ring known as a quoit, which is attached to a rescue line and intended to be thrown to people outside the raft who need to be pulled in. Next to the survival display were tacked up dozens of press clippings about boats fined for various environmental infractions—and about sunken fishing vessels and their survivors and victims.

The primary focus of the course was to teach the students the types of fish they would be working with in Alaska and how to sample the catch. Many hours were also spent reviewing the different types of boats and fishing gear they might encounter, which would influence how often they would sample and in what manner. The students were exposed to an overview of the complex laws that govern the fisheries, and were spoken to by a NOAA enforcement officer who explained how to report environmental violations.

Jay had hours of homework most nights. There was a midterm and a final and the students had to score at least 80 percent on each to pass the class, and get eight out of ten species on a fish identification test as well. In a separate classroom a floor below, the instructors pulled a few dozen frostbitten specimens from the freezer and laid them out on stainless steel tables. The samples were old and had scars from years of poking and prodding by nervous students. To Jay, the fish test was the hardest part of the class. Each student had to use an identification key to pinpoint the exact species. It was harder than it sounded. Most of the samples looked very different from the pictures in Jay's book, and even once you narrowed it down, there were so many varieties of each fish: five types of salmon, fifteen skates, sixteen sculpins, more than thirty kinds of rockfish.

In the last week of class, the students were split into two groups: men and women. The women gathered around a veteran female observer and were given a chance to ask questions about how to handle potential sexual harassment on a boat. They were drilled with some time-earned wisdom: You'll be in rubber boots and a slimy sweatshirt, your hair pulled back in a ponytail, and smelling like fish guts. You won't have looked in a mirror in weeks, but you'll still be the most beautiful woman the fishermen have seen in months. They'll probably let you know it. The women were advised to maintain professional boundaries and—above all—not to get sexually involved with a fisherman on a boat. It would damage their reputation and make it harder to do their jobs. Don't kid yourself into thinking you'll just keep it secret, the new observers were told. Gossip spreads fast on a fishing boat.

Meanwhile, the men got a lecture in how to avoid coming off as smart-ass college boys.

The last week of class, the entire group loaded into a bus for the ride to a local university's indoor pool, where Jay and his classmates practiced getting into their survival suits. They were timed to do it in under a minute—considered the industry standard. They were taught a very specific method: Lay the suit out on the floor, unzip it, and sit down on top of the open torso to wiggle your feet into the neoprene legs. Once their legs were in, they were to kneel forward and pull on the upper half of the suit.

The weak arm went in first; then they should use their strong arm (the right for right-handed people) to secure the suit's hood over their head *before* sliding in the second arm and zipping the suit up over their chins with the long string attached to the plastic zipper.

Back in the classroom, the instructors provided a few tips for getting on suits quickly and effectively. Plastic bags can be stored in the legs of the suit and slipped over the shoes to help the feet slide in more easily. Hats should be removed and long hair pulled back to be sure the seal between hood and face is watertight. If it's time to put on the suit, it's time to zip it up all the way. There were plenty of instances of bodies being pulled out of the water with a suit on and the zipper opened up to the sea.

The observers-in-training each shuffled up to the pool's diving board, crossed one leg over the other, held one hand over their face to prevent their mouth flap from being forced open, placed the other on their head to hold the hood in place, and jumped into the water. They swam laps in the suits—always on their backs, to prevent water from leaking into the neck. And they practiced climbing into a life raft deployed in the pool. They were taught how to pull their upper body onto the edge of the raft, and then kick their feet hard to help lift their lower body up in the water. From that position, it was easier to be pulled into the protective structure. Each trainee took a turn at pulling and being pulled.

The observers practiced linking up in the water in small groups, rafting together head to feet, or in a chain, with each person's legs wrapped around the torso of the person in front of them. Finally, the whole group linked arms to create a large, pinwheel-like circle. With their heads toward the center and their feet facing out, all the students could kick at once, a maneuver that would theoretically allow a group of people to be more easily spotted in the water from an aircraft flying far overhead. There was some additional warmth to be gained from staying together; there was a definite benefit regarding visibility. Perhaps the greatest advantage, though, was to morale. Feeling like you're helping someone else to survive can sometimes be the key to your own survival. The instructors repeated the same points again and again. Most of them had experienced a few close calls out there themselves. They tried their best to drive the safety lessons home.

After the pool day, back at the training center, the students picked out their own personal Gumby suits from several sizes and brands—eight different fits in all. Jay found a good match, then rolled the suit back up and stuffed it into its storage bag along with a couple pieces of hard wax he'd been given to keep his zipper working smoothly.

He completed the program on January 15. The next day he was on a plane. His first vessel was a small catcher boat out of Akutan, a village of eight hundred people thirty-five miles east of Dutch Harbor. The *Ranger* was his second ship. He'd been aboard for about three weeks now.

WHEN GWEN ARRIVED ON MARCH 19, she and Jay decided she'd take over Chris's noon-to-midnight shift. As the junior observer, Jay would stay on the midnight-to-noon. So far there hadn't been much work. The day Gwen got on the *Ranger* for the first time, the ship left Dutch Harbor to fish for yellowfin sole. They'd barely done any fishing at all—just four hauls—before the fish master ordered the boat back to Dutch. Word was that the other FCA boats were doing the same. They'd given up on yellowfin fishing for now; instead they'd be steaming several hundred miles west to fish for Atka mackerel.

They set out at noon on Saturday, March 22. As always, Gwen recorded the time in her logbook. She spent much of the afternoon in the wheelhouse. She watched a movie in her cabin and was asleep by 10:00 P.M. She woke up to the A-phone ringing.

Jay picked up.

"Hello? Hello?"

No answer.

As soon as he put the receiver down, the phone was ringing again.

Gwen worried that Silveira might be trying to reach them from the wheelhouse. Maybe he had a question about paperwork. Gwen knew both Silveira and Captain Pete were new to trawling. They'd been having trouble with some of the record-keeping required by recent changes in fishing laws. She got up.

The ship's alarm went off just as Gwen climbed the last stairs to the wheelhouse. Silveira was inside.

"This is bad. It's really, really bad," he said.

"What's going on?" Gwen asked.

"We're flooding."

Silveira was in the middle of a conversation with the ship's assistant engineer, Rodney Lundy. Moments later, Rodney's boss, Dan Cook, came through the door.

Gwen sat down and listened.

It was the rudder room, Rodney reported. The water was already thigh-high.

The engineer thought it was already too late to stop the flooding. They had to focus on blocking it, partitioning off the water and, above all, preventing it from spreading to the engine room.

GWEN HAD A PERSONAL LOCATOR BEACON, or PLB, essentially a hand-held EPIRB. Every Alaskan fisheries observer is issued their own beacon. The several-hundred-dollar gadget is registered to a particular user. Once a beacon is transmitting, a NOAA station in Maryland picks up the signal instantly. They get a GPS hit that's accurate within a few dozen yards and should immediately contact the observer's employer and local rescue authorities.

Each time she got on a new boat, Gwen looked for the best place to store her survival suit and sandwich-size PLB. On the *Ranger,* she had her beacon on a hook in her cabin; her suit was stored in the wheelhouse, separate from the suits for the crew. Jay showed up in the wheelhouse soon after the general alarm started going off. When he realized it was a real emergency, he ran back down to the observer cabin to get his PLB and Gwen's. Meanwhile, Gwen grabbed her survival suit and started pulling it on. She set off the beacon as soon as she was suited up. Jay set his off at about the same time.

Gwen was still in the wheelhouse when the satellite phone call came in. Silveira picked up. It was the representative at Saltwater, calling for Jay. Was there a real emergency?

"Yes, Jay did activate his PLB," Silveira reported. "We're flooding."

Silveira was looking right at Gwen as he hung up. "Do you have yours activated?" he asked her.

"Yes, I turned it on!" Gwen replied. "They didn't ask anything about me?"

"No," Silveira said. "They didn't."

FOR MORE THAN AN HOUR, GWEN LISTENED as Silveira and Captain Pete consulted with the engineers, sent men down to try to control the flooding, and made calls to the other FCA trawlers and the Coast Guard. Everyone else had gotten into their suits and had been ordered back out onto the deck. Gwen figured she could stay inside if she just stayed quiet and out of the way.

From the moment she walked into the wheelhouse, Gwen had known the situation was dire from the look on Silveira's face and by his tone of voice. Based on what she overheard, she didn't think the *Ranger* could be saved, but she did think it was possible to contain the flooding until another ship arrived. She imagined abandoning ship onto the *Warrior* or, better, onto a Coast Guard vessel.

Then she overheard Silveira and the engineers discussing the possibility of the *Warrior* towing the *Ranger* back to Dutch Harbor.

"We're not going to get towed in," the assistant engineer, Rodney Lundy, said. "There's already water spraying in around the doors."

I'm getting into a life raft, Gwen thought. I'm going to have to get into a tiny life raft in the middle of the Bering Sea.

Gwen thought about her kids. She always called home when she was in port between trips. But this time, she'd only talked to them for a couple minutes. She hadn't seen them in two months.

What am I doing here? she thought.

About twenty minutes after Gwen overheard the engineers talking about the spraying water, the ship's lights began flickering.

"The water's made it to the engine room," Gwen heard Cook say. The engines were sputtering, gurgling.

The lights went out.

Silveira was at the helm. "I've lost steering," he said. He repeated the words several times.

"The engines are backing up," Silveira said. "They're backing up!"

Moments later, Silveira gave the order to abandon ship.

Gwen was assigned to the number two life raft, stored on the

Ranger's port side. When she got out on deck, the emergency squad member who was supposed to launch that raft didn't know how. The Jacob's Ladder wasn't tied off. Gwen felt panicked. She knew how to launch the raft. At her last briefing in Seattle, there had been a refresher class on safety training; a Coast Guard officer had come to talk to the group and each student had simulated launching a life raft. She had no idea, though, how to tie off the Jacob's Ladder.

Gwen was relieved when Evan Holmes ran over from his post on the starboard side of the ship and tied the raft's port ladder to the boat. He saw Gwen standing by the rail.

"Don't you worry. We're going to get you off this boat," Evan yelled to the fisheries observer.

The words of reassurance made Gwen feel calmer. She looked at the raft.

"I just pull the pelican hook, right?" she yelled back to Silveira over the growling engine noise.

What if I do it wrong? Gwen thought. She was seized with fear. This isn't just my raft, it's everybody else's raft, too. She knew that was the correct way to launch the raft, but she wanted the confirmation of the mate.

"Yes, pull the pelican hook!" Silveira yelled from the wheelhouse.

Gwen and another crewman pulled the hook and pushed the raft over the port rail. The bulk hit the water, then nothing. Gwen couldn't see it. She had no idea if it hadn't inflated at all, or if it had just somehow swung out of sight. The ship was listed far to starboard and Gwen's side of the boat was raised high above the water.

Maybe the raft is stuck down against the hull, where we can't see it, Gwen thought.

She could see the painter line. It was right there in front of her, pulled taut. The raft had to be somewhere. Then, not much more than a minute after they'd launched the raft, she saw the line snap.

Gwen strained her eyes into the blackness. Still, she couldn't see the raft.

Gwen could hear people yelling "Abandon ship!" but jumping blindly seemed like a bad idea. Her training had taught her to get directly from the vessel into a raft if humanly possible. The instructions had been repeated again and again: Get into the raft. You *must* get into the raft in the Bering Sea. The words came into her mind now.

Gwen went back into the wheelhouse to talk to Silveira. Pete was inside too, along with the fish master and the engineers. The two head officers were taking turns working the radios.

It was 4:15 A.M. when the officers on the *Alaska Ranger* radioed the Coast Guard that they had lost life rafts.

"COMMSTA Kodiak, COMMSTA Kodiak, this is the *Alaska Ranger.*"

"*Alaska Ranger,* this is COMMSTA Kodiak, over," watchstander David Seidl answered.

The ship's transmission was completely muddled by static.

"*Alaska Ranger,* this is COMMSTA Kodiak," Seidl said once more. "Request you say your last again, over."

"Yeah, the boat took a big-ass list to starboard," the ship's officer answered. "We launched the port raft. The painter broke." There was more, but again, static drowned out the transmission. After a few minutes, another few words were audible: "We're getting into the rafts right now."

"*Alaska Ranger,* COMMSTA Kodiak, roger," the reply came back. "Have you weak or readable. I think I got a good copy. Understand your vessel listed to starboard, and you lost a couple rafts into the water. Is this correct, Captain?"

"Yes, we did, we lost one of them. On the port side. The, uh, the painter broke."

"Roger, Captain. Good copy. Be advised we have an ETA for Coast Guard rescue 60, should be there in approximately five, zero minutes. How copy on that, sir?"

"I copy."

"Roger, Captain. We appreciate the information. Please keep us informed. We'll be standing by for you, sir."

BETWEEN CALLS, PETE TURNED TO GWEN. "This vessel is going to capsize any minute," the captain told her. "You have to get off. *Now.*"

Gwen couldn't mistake the urgency in Captain Pete's voice.

She went back outside, to the stern deck on the starboard side. One of the other rafts was floating not too far from the ship. It looked pretty stable in the water. A group of fishermen were standing around an empty life raft cradle.

"Go, go, go, go!" Gwen yelled. "We have to get off this boat now!"

Gwen watched as one guy grabbed hold of the painter line and shimmied down into the water. Gwen followed, but almost immediately lost her grip on the rope.

She bobbed up, spitting out a mouthful of salt water. She could see the raft. She started swimming: two breaststrokes. Then she stopped. She thought back to her training. Stop and think, she'd been instructed back in Anchorage. Try to relax. Get your breathing under control. If you have to swim, swim on your back, or you'll end up with water inside your suit.

All I have to do is get hold of the painter line, Gwen thought. Then she could exert less energy by pulling herself into the raft. She was lying on her back in the water with her legs and arms crossed, the best way to conserve heat, she remembered, when she saw the line. She reached out and grabbed it through the neoprene mitts of her survival suit. Hand over hand, she walked her way up the rope to the edge of the bobbing life raft.

Through the open door of the tented shelter, she could see that a couple of the Japanese technicians were already inside. As she reached the shelter, one of the men leaned out, and grabbed on to her. With one strong yank, he hauled her up into the safety of the tented compartment.

THE CREW OF THE COAST GUARD'S Hercules C-130, rescue plane 1705, launched in textbook time, half an hour after pilot Tommy Wallin got the call. From Anchorage, it was about nine hundred miles to the sinking ship. They should be on scene within three hours.

The Herc had been airborne for less than an hour when a communication came over the radio. The boat was sinking. People were abandoning ship, straight into the water.

The mood in the aircraft grew tense. Pilots Wallin and Matt Duben had already calculated their route to the *Ranger,* but now they decided to ascend another few thousand feet. The airframe can get more speed at higher altitude—and flying higher also helps preserve fuel. The 60 Jayhawk helicopter from St. Paul Island would almost certainly reach the spot before they did, but once they got to the emergency site, the C-130 crew would take over on-scene communications. At their altitude, the plane would have the ability to communicate directly with the 60 Jayhawk, the Coast Guard cutter *Munro,* the base in Kodiak, and District Command in Juneau. Until they were on scene, the Coast Guard assets would be communicating as best they could through VHF and UHF transmissions. Until they got there, the crew of the C-130 wouldn't be doing anybody any good at all.

BACK AT THE COMMUNICATIONS STATION in Kodiak, the transmissions from the *Ranger*'s officers were growing more and more muddled.

"*Alaska Ranger, Alaska Ranger,* this is COMMSTA. Be advised I have you weak and barely readable," watchstander David Seidl radioed to the ship at 4:22 A.M. "Understand you have lost *all* your power, over?"

"That is correct. I also have a twenty-five-to-thirty-degree starboard list. I got two people, two people, in the water."

"*Alaska Ranger,* this is COMMSTA. Confirm, two people have gone *in* the water at this time, over."

"Roger, I've got two people in the water," the response came back from the ship. "I have no power, all right?"

"*Alaska Ranger,* this is COMMSTA. Do you have a visual on the two people, over."

"No, I do not at this time."

"*Alaska Ranger,* this is COMMSTA. Roger, understand. Stand by one, over."

"Roger."

Several minutes later, Seidl attempted to make contact with the *Ranger* again: "*Alaska Ranger,* this is COMMSTA Kodiak, over." He waited.

Nothing.

"*Alaska Ranger,* this is COMMSTA Kodiak." Still, silence.

At 4:36 A.M., almost two hours after David Seidl heard the first "Mayday" broadcast into his Kodiak cubicle, the watchstander scrawled the words "No Joy" in his notebook.

He had lost all communications with the sinking ship.

Chapter Seven

Alone in the Waves

Ryan Shuck stood at the starboard rail, near the empty canister that had held the number three life raft. The guys in charge of his muster group, Evan Holmes and Eric Haynes, were telling people to go, that they had to get off the boat and try to swim for the raft.

Indio Sol, a Thai crewman everyone called by his nickname, "Rasta," went first.

"I guess I'm going in," he said.

Then just like that he climbed over the rail and descended the Jacob's Ladder into the water. A young processor named Kenny Smith went next.

Ryan watched each man hit the waves and take off—two red dots drifting fast toward the boat's bow. The two starboard life rafts were tethered to the moving ship with their painter lines. The lines were pulled taut and the rafts were a good distance beyond the bow. Ryan watched as his two crewmates drifted past the end of the ship, then beyond the rafts. He couldn't tell if his friends saw the life rafts—or if they were even trying to swim at all. They were already just tiny specks, powerless under the strength of the waves.

Ryan climbed down onto the ladder and tried to launch himself farther away from the side of the ship. He surfaced quickly and started swimming on his stomach, pushing hard for the nearest raft. It seemed like his strategy was working. The raft was ten feet away, then three. He was there. He hit the dead center of the tented structure and tried to grab on. But with his hands wrapped in the thick neoprene he couldn't get a good hold on the slick rubber raft. It was like trying to grab and climb onto a giant inner tube that was rushing by in white water. Ryan was up against the side of the raft. Then he was sucked underneath it. He couldn't see. He couldn't breathe. And then he surfaced—with the raft behind him.

ERIC HAYNES BALANCED ON THE EDGE of the deck, staring down toward the waves. He could see the two starboard-side life rafts bobbing out beyond the *Ranger*'s bow. Then both rafts seemed to disappear. It was so chaotic that it was hard to tell what was happening. Most people were already off the ship, but Eric didn't know if any of them had made it to the rafts. Then all of a sudden the rafts were in view again. Eric was shocked to see that they'd somehow swung all the way around the boat and were bolting back up the starboard side. The *Ranger* was still at a severe list. The drop from deck to water that normally would have been a fifteen-foot plunge was only a few feet. It felt almost like stepping off the end of a dock into the water.

Eric sank under and swallowed a mouthful of seawater.

When he bobbed back up, he saw the raft's painter line—which had broken off from the ship—right in front of him. He reached out, wrapped his hand around the line, and was immediately jerked under. He fought his way up and dragged himself along the line to the raft. He'd almost reached the shelter when he saw someone a few yards away. The man was floating spread-eagled, facedown in the water.

Eric reached the raft and then maneuvered around it, pulling himself closer to the floating body using the ropes built into the sides of the shelter. He grabbed on to the floating man's leg and turned his head up out of the water. It was Joshua Esa, a processor from Anchorage.

Eric couldn't tell if Joshua was alive as he struggled to pull him back around to the raft's entrance. He wasn't responsive.

I just need to get him in the raft, Eric told himself. *If I can't, I'll just have to stay with him in the water.*

Eric pulled Joshua around to the entrance of the shelter. David Hull was in the opening. He'd have to get himself in first, Eric realized. With David's help, Eric kicked and pulled his way in, finally collapsing like a huge gaffed fish on the soft floor of the raft.

It took a minute for Eric to pull himself up from the raft's floor. Every time he moved, he seemed to sink deeper into the collapsible plastic. By the time Eric got back to the raft's entrance, David had lost his grip on Joshua, who was now drifting away.

Then Eric saw Joshua kicking his feet. He was alive.

Seconds later, Boatswain Chris Cossich floated into view next to Joshua. Chris grabbed onto Joshua, and within seconds the two were at the door of the raft. Eric yelled for the other men in the raft to help, but the five or six people inside just sat there. *Maybe they're in shock*, Eric thought as he grasped onto Joshua and tried to pull him into the shelter.

It was no use. Without Joshua's cooperation, it was like trying to lift a bag of cement over a high railing. Then Eric had an idea. He yelled to Chris to lift Joshua's legs into the doorway. Eric leaned out of the raft and hooked one arm under Joshua's knee and the other around Joshua's shoulder. He'd pull while Chris pushed. They struggled for a few moments, and then Eric and Joshua tumbled back into the raft. Chris and a few other guys climbed in after.

After catching his breath, Eric stared out the open doorway. He couldn't see anyone else. Lots of guys didn't make it into the rafts, he thought. They're out there, and we have to look for them.

"This is what we're going to do," Eric told the group. "We'll take turns keeping a lookout for those guys. We need to keep yelling to let them know we're out here."

For more than an hour they screamed almost constantly. Nobody yelled back. It seemed like the weather was getting worse. With the door open, water splashed inside the raft, making people even colder. Joshua was awake, but he didn't look good. He was zoned out. Several of the guys were shivering. Finally, Eric zipped up the door and leaned back against the side of the jolting raft.

Factory manager Evan Holmes was bobbing in the swells. Back on the ship, he'd been standing next to a friend of his, a small-framed Laotian guy named Phouthone Thanphilom whom everyone called P. Ton. The men were holding on to each other as they balanced on the listing deck.

"Hey, Holmes, you'll take care of me, right?" P. Ton asked Evan back on the boat.

Evan thought of P. Ton as his little buddy on the ship.

"Yeah, I'll take care of you," he said.

He was relieved when they found each other again after just a few minutes in the water. Evan could immediately see that P. Ton was panicking, trying to swim on his stomach in his enormous suit. The Laotian man couldn't have weighed much more than 100 pounds.

"Get on your back!" Evan yelled as he helped to roll P. Ton over in the waves. He noticed that his friend's strobe light was off, and turned it on.

"Grab my legs," Evan told P. Ton as they linked together in the swells.

KENNY SMITH FELT LIKE HE'D BEEN in the water for a long time. He had been one of the first to jump off the boat, and was carried right past the raft. For a while after he hit, he felt all right. It was probably the adrenaline keeping him warm. But water was slowly leaking into his suit. The fishing gig was pretty much Kenny's first real job. He was twenty-two and had been on the *Ranger* for about nine months. Before Alaska, he'd worked as a newspaper delivery boy. And he'd spent some time in jail after the police found stolen goods in his apartment. He was storing the stuff for a friend. Just a few weeks before, he'd called his girlfriend back home in eastern Washington. It was her birthday, and it cost him thirty bucks for a twenty-minute call on the ship's SAT phone. She told Kenny she was pregnant.

By half an hour after he abandoned ship, Kenny was freezing. The waves were crashing right over his head. He felt like the water in his suit was dragging him down. He didn't want to swim anymore. He felt like he couldn't hang on. *Screw it, I'm going to die,* Kenny thought.

* * *

Coast Guard Jayhawk pilots Brian McLaughlin and Steve Bonn stared out into the dark night. It was flurrying on and off, with wind gusts up to 35 miles per hour. The only thing between St. Paul and the spot they'd programmed into the aircraft's computer was the inky blackness of open ocean.

The time spent approaching a rescue scene is a chance for an aircrew to talk out scenarios—to discuss what they might find, what actions they'll feel comfortable with, and how much risk they're willing to take. The Jayhawk crew had already heard from the Coast Guard cutter *Munro* that at least two fishermen from the *Alaska Ranger* had gone overboard without getting into a life raft. It sounded like the flooding was advancing fast.

The aircrew hashed out their options. Obviously, they couldn't airlift a forty-seven-person crew in one load. If the ship couldn't be saved, they'd most likely be making multiple trips. Even half a dozen people would be cramped in the helo's cabin, especially if they were wearing Gumby suits. If necessary, though, they might be able to squeeze in twice that many.

The men had calculated the distance from the *Ranger*'s last known position to the Dutch Harbor airport, the nearest landing spot. Given the poor weather conditions, they'd have to circle around the north coast of the island and approach the airport from the east, which would make the total trip close to 150 miles. If they needed to return to the sinking site, the crew agreed, it would make the most sense to lower the survivors to the *Munro*.

McLaughlin had been communicating with Operations Specialist Erin Lopez. Now he heard another voice break through over the radio.

"Rescue 6007, this is cutter *Munro*." It was the ship's commanding officer, Captain Craig Lloyd.

"Sir, we think we'd like to bring any survivors straight to you," the pilot reported.

"That seems like the best plan," Lloyd agreed. "We'll be ready to work with you in any way necessary."

Steve Bonn was in the right seat, flying the helo. The thirty-nine-year-old pilot had just a few months left in his Alaska tour. In early summer, he was scheduled to transfer down to Elizabeth City, North Carolina. Earlier in the week, McLaughlin had decided that if his crew got a search and rescue case during this St. Paul deployment, Bonn would take the reins. McLaughlin wanted to give Bonn the chance to go out with a bang. He certainly hadn't imagined anything like this—though he was more than happy to have the more experienced pilot in the right seat. Like more than a few Coast Guard pilots, Bonn was former Army. For ten years, he'd flown the military's Blackhawk, the platform from which the Coast Guard's search and rescue Jayhawk had been designed. "Sikorsky made the 60 to get shot at and keep on flying," Jayhawk pilots often boasted of their aircraft. The same was said of the Army pilots: They could take just about anything.

About fifty miles out from the sinking site, McLaughlin was able to reach the *Alaska Ranger* on the VHF radio. First Mate David Silveira told him the situation had deteriorated significantly in the last half hour. Only seven people were still on board, and the 184-foot fishing vessel was listing to 45 degrees. It looked like it might capsize at any minute. Most of the crew had already abandoned ship, the officer said. Some of them had made it into life rafts. Others hadn't. He didn't know how many.

FOR A FEW MINUTES AFTER HE was sucked under the life raft, Ryan Shuck struggled against the breaking swells, trying to make it back to the circular shelter. But it was pointless. He was too far away and already exhausted. He lay back in the water, letting his head rest against the inflatable pillow at the neck of his survival suit. His heart was pounding.

Ryan tried to concentrate on how his suit supported him in the water and how best to avoid being pummeled by the swells. He did his best to position himself with his back to the breaking waves. He looked up at the moon, skipping in and out of view in the black sky. In the distance, he could hear someone yelling: "I can't swim, I can't swim. I don't know what to do!"

Ryan tried to talk himself into calming down.

Every time he rose up on a crest, he could see lights spread out behind him in the water. It seemed like he was farther downwind than anyone else. There was a small cluster of lights about two hundred yards away. For a few minutes, Ryan tried to swim toward it, but the waves kept turning him around. He couldn't even keep the lights in sight, with the way the water was flipping him around. He decided it would be better if he just stayed still.

Gazing back toward the ship, Ryan could see at least half a dozen tiny, solitary beacons flickering among the waves. There was just enough moonlight to make out the outline of the *Alaska Ranger* bulging from the ocean. The ship was dark, just a shadow, really. Ryan watched as her bow turned slowly up, finally pointing straight toward the sky. The wheelhouse was at the waterline when, eerily, the lights inside flickered on for a moment.

There's still some power, Ryan thought. Maybe she'll right herself. But then, in a matter of seconds, the ship plunged straight down, swallowed whole by the dark sea.

IT HAD TAKEN RYAN TWO YEARS to get a job on a fishing boat—two years working even crappier jobs for much crappier money.

He'd grown up in Libby, Montana. When he was thirteen, his family moved to Juneau, Alaska's tiny capital city. His dad had heard about work at a new mine up there. Ryan went to middle school in Juneau while his dad hauled rubble out of the excavation site. But it was only a year before the work dried up, and the family packed for the ride back to Montana.

After high school Ryan worked in logging for a while. Eventually he found himself in Great Falls, where he got a job cleaning exhaust hoods for restaurants. He was traveling all over the West, making $12 an hour—not bad for Montana. But he kept thinking about Alaska.

In 2005, he saw an ad in the Great Falls newspaper. He went to a meeting at the local Best Western. The cannery hired pretty much everybody, provided you passed the drug and alcohol test. Ryan got the job and was soon in Dutch Harbor, working twelve hours a day on an assembly line in a fish processing plant. They did everything: crab, cod, pollack. He tried to talk to people about getting on a fishing boat, but those jobs were tough to get if you didn't have experience. And it was impossible to get experience if you couldn't get on a boat. The next year, he got another cannery job, this time in Kodiak. In the late spring of 2007, he started making phone calls to Seattle fishing companies.

Ryan had seen a lot of little boats at the Kodiak docks. He hoped to get on one of the bigger ships, which he knew mostly had owners in Washington State. He thought a larger boat would be a smoother ride and a little nicer to work on. For the most part, Ryan got answering machines. But at the Fishing Company of Alaska, someone picked up. He was told they could put him on a boat if he could get himself to Seattle by Friday. It was Tuesday. Ryan bought the ticket and showed up at the FCA's Seattle office a couple days later.

He filled out some paperwork and watched a video of processors working in the factory of an FCA boat and of a ship under way in rough seas. The company didn't try to sugarcoat it. He'd probably get seasick. He'd definitely be sleep-deprived. He'd most likely dream about fish, about kicking fish and slicing the heads off fish and wading day after day after day in a river of thousands of dead or almost-dead fish. If you don't think you can handle it, leave now, the recruiter told Ryan and the other hopefuls in the conference room. It's not too late to back out. The job's not for everybody. No one will think less of you if you don't want to do it.

There was no chance in hell Ryan was backing out. When the orientation was over, he walked to a nearby health clinic and pissed in a cup. A few hours later, he got a phone call: Be at the Seattle airport at 6:30 the next morning. The ticket would be waiting.

The job sucked, and so did the pay: fifty bucks a day to start. But turnover was high. A bunch of guys quit after his first trip on the *Alaska Ranger.* Ryan didn't. He got promoted to tallyman, one of three men on board responsible for keeping track of the number of cases of fish loaded into the freezer during each six-hour shift. The new job came with a $30 a day raise, plus an offload bonus of five cents per case. Ryan had only been working for the company for two weeks.

Back in the fish-processing plants, Ryan had heard from the older guys that if you're still young, you're better off working on the ships, where there's more money to be made—provided you're strong enough to do the work. Ryan was making more money. But the work was unpleasant and dangerous. It wasn't uncommon for men to lose fingers. The catch was sometimes hauled up covered with muck and he'd seen fish accidentally coated with hydraulic oil processed and frozen along with the rest. The job made him never want to eat fish again.

He always felt ready to go home. In between stints in Alaska he went back to Spokane, where he'd moved from Great Falls with his girlfriend, Kami, soon after he got that first cannery job. He'd be so tired, so sick of fish. The smell would have coated all his clothes and penetrated his skin. After a few months in Alaska, his hands were swollen and chapped from being in salt water all day. But once Ryan was back at home, he'd think about the boat and his friends from the *Ranger*. Sometimes they talked on the phone, retelling funny stories from the fishing season. They mostly remembered the good stuff. The practical jokes, the nights in the bar. It was never, "Hey, remember when your back hurt so bad you couldn't get your boots on?"

The first couple of times, Ryan felt like going back up to Alaska when the time came around. After a couple months at home, he and Kami often had trouble getting along. Even she said it was better for their relationship when they spent time apart. He liked intense jobs. All the free time made him restless. Sometimes he drank too much and got into fights. But when Ryan got back to Spokane in late 2007 he felt like things were going to be different. He had a lead on another job, cleaning restaurant exhaust hoods again, this time in Washington State. He said no when the FCA called in December, asking him if he'd come back for the winter A season. He thought he was done with Alaska and with fishing. But on January 2, there was another call with an offer of more money. The FCA didn't have many experienced guys coming back to the *Ranger*. They said they needed him. Ryan's other job still wasn't a done deal. A few days later, he was on the plane. One more three-month contract in Alaska, Ryan figured. Then he'd call it good.

LIKE MOST NEW PROCESSORS, RYAN DIDN'T ask questions about the company's safety record. His priority was getting a job—

and a paycheck. In fact, his boat would be the second ship lost by the Fishing Company of Alaska.

Ten years earlier, on February, 11, 1998, a 198-foot long-liner named the *Alaska 1* sank after colliding with a freighter thirty miles north of Dutch Harbor. It wasn't a fair fight. The container ship, the Korean-operated, Panamanian-flagged *Hanjin Barcelona,* was almost five times the length of the *Alaska 1.* After the collision, against the wishes of local authorities, the foreign vessel continued on its planned route toward Taiwan. If any repairs were necessary, the ship's crew evidently felt confident that they could cross a few thousand miles of Pacific Ocean before tending to them.

No one was seriously injured in the incident. Most of the *Alaska 1*'s crew of thirty-three were sleeping when the collision occurred, at close to 11:00 P.M. All of them abandoned ship into two life rafts and were quickly rescued by a Good Samaritan vessel. The successful evacuation was seen as a validation of the 1988 law that required safety gear and training on commercial fishing boats, though there were a couple of serious fumbles. One large crew member couldn't fit into the only remaining survival suit and ended up abandoning ship without a suit. And once the rafts were in the water, the crew couldn't find the cutters supplied to sever the painter lines and free the rafts from the boat. Luckily, there was enough time for a crew member still on board to run down to the ship's store and grab a couple of knives to cut the lines.

The *Alaska 1* was at fault in the crash. When the courses of two vessels cross each other and there is a risk of collision, the ship that has the other on its port side has the right of way. In this case, that ship was the *Hanjin Barcelona.* However, the Coast Guard investigation concluded more was at play than just a confused traffic rule. A preliminary drug test administered soon after the sinking indicated that the *Alaska 1*'s on-duty officer, First Mate Randy McFarland, had cocaine in his system at the time of the crash. He'd also had just eight total hours of fragmented sleep in the previous thirty-six to forty-eight hours. Due to a problem with the size of McFarland's urine sample, the Coast Guard was ultimately unable to include the drug test results in evidence. Several months later, though, McFarland was arrested in Seward, Alaska, on charges of selling cocaine. (He served time in jail, and now works for a sports fishing company that caters to tourists on Alaska's Kenai Peninsula and runs a local lodge named the Fish Whisperer.)

The *Alaska 1,* meanwhile, was never salvaged from the ocean floor. It took less than an hour for the vessel to disappear below the surface and sink to the bottom, in eighteen hundred feet of water.

Just a few years before, the FCA had suffered still another major casualty—this one far more tragic. On May 27, 1995, the trawler *Alaska Spirit* was moored near a dock in Seward when a fire that began in a stateroom burned much of the interior of the boat and killed the ship's captain.

A National Transportation Safety Board (NTSB) investigation ultimately determined that the fire was most likely started by an electric rice cooker in a room normally inhabited by the boat's assistant fish master. The ship had no sprinkler system and no smoke detectors. Its fire hoses were incompatible with available hydrants, and the crew had little firefighting training.

The blaze began about 2:00 A.M. and wasn't entirely extinguished until 11:00 the next morning. An autopsy determined that the captain, who was overcome by the fire while asleep in his stateroom, was intoxicated at the time of the incident. The damage to the ship was estimated at $3 million.

The NTSB concluded that the lack of fire-safety standards for commercial fishing vessels contributed to the damage and loss of life on the *Spirit*. The incident prompted the board to issue a series of recommendations on improving fire safety in the fishing fleet. None of them, however, ever resulted in a change of law.

RYAN'S MIND WAS RACING. He was pretty sure he'd heard the ship's officers talking to the *Warrior*. The other FCA boat would be on its way, but would they have relayed a message to the Coast Guard? If they did, the Coasties would be coming all the way from Kodiak, Ryan thought. That was so far, hundreds of miles away. It was still dark. Would it be easier to spot him at night, or during the day? he wondered. Maybe at night. His strobe light was still flashing. He thought about how he'd look from the sky. Would they see the suit, the light? Jesus Christ, how long could he stay like this?

The waves were huge: twenty-footers, Ryan guessed. From the deck of the ship, seas this size wouldn't be any sort of spectacle. It was a different story when you were submerged in the goddamn things. He couldn't keep the freezing spray out of his nose and eyes, the only parts of his body that weren't protected by the suit. What if no one was coming?

Ryan thought about unzipping his suit. He could lie there and freeze to death, or he could make it quick. Just get it over with.

The water was 35°F and the air temperature about 15°F. Even in water as warm as 75°F, the summertime temperature in Hawaii, immersion hypothermia is common after prolonged exposure. Body heat transfer is one hundred times greater in water than in air of the same temperature, according to *Wilderness Medicine*, the academic tome of survival in extreme conditions. Ocean temperature has to reach the low nineties before a naked person in the water has neutral heat loss. In cold seas, with little protective clothing, it takes just a couple of minutes for the first stages of hypothermia to set in.

When a person first hits frigid water, his or her body has a "cold-shock" response. The colder the water and the more sudden the exposure, the more extreme the response. Survival experts advise that, if possible, it is better to enter cold water gradually, rather than in a sudden, full-body plunge. The rapid cooling of skin that comes with sudden immersion can cause a gasp reflex, which itself can cause drowning, especially in rough seas. It also causes a tendency to hyperventilate. Normally, that hyperventilation will begin to diminish within seconds, though extreme emotional stress or panic can cause it to increase instead. Uncontrolled hyperventilation can lead to muscle weakness, numbness, and even fainting—all of which can lead to drowning. Some cold-water immersion victims don't survive the first two minutes.

Survival school instructors often teach the 1-10-1 rule of cold water immersion. You have one minute after you're submerged to gain control of your breathing: That's the first "1." Don't immediately start struggling to get out or swim to safety. Instead, focus only on your breathing for those first sixty seconds. Gain control by taking slow, deep, conscious breaths. Think to yourself: I won't panic.

Now you have approximately ten minutes before the cooling of your extremities will seriously impact mobility, especially in the hands, where blood circulation is negligible. Finger stiffness, loss of coordination, and compromised motor control will soon make it difficult, if not impossible, to carry out survival efforts, such as grasping a rescue line, according to Alan Steinman, M.D., an expert in cold-water survival and a retired Coast Guard Admiral. Do you have any survival tools (whistle, flares, strobe lights) that will help rescuers find you in the water? Make sure they're operating and accessible now.

Survival instructors advise using these ten minutes to take stock of your situation and attempt to improve it. Hoods should be raised and drawstrings tightened, which will increase insulation and reduce water circulation. Do everything you can to keep your head out of the water. Is there a life raft or other form of flotation close by? If so, expend the strength to reach it now. Being inside a life raft is always better than being exposed in the open ocean. Absent a raft, any sort of flotation that can be used to prop at least some of your body out of the water—buoys, life rings, coolers, or other debris—should be used. It may not seem like it will make a difference, but even a small amount of flotation is better than nothing. The more of your body is out of the water, the better your chances for survival.

If no flotation is available, survival experts recommend a position known as HELP (for heat escape lessening position). Assume a fetal position in the water, with your arms pulled up against your chest and your legs raised up, and pressed together. Heat loss is highest from the groin, the lower torso, and the neck—areas of the body with a relatively thin layer of soft tissue and a relatively high rate of blood flow. If possible, huddling together with other survivors is also a good survival strategy. It will help retain heat and often improves morale.

Almost all cold-water immersion victims who do not have survival suits lose consciousness within an hour. This is the final "1" of the "1-10-1." That's loss of consciousness—not death. Even if you aren't rescued within an hour, you can increase your chances of survival by trying to position your head so that it doesn't fall underwater when you pass out.

There are many documented cases of victims being revived even after losing all vital signs, a phenomenon that's more common in cold water than warm (and more common among children than adults). A cold-water near-drowning victim may

The *Alaska Ranger* was one of seven ships owned by the Fishing Company of Alaska (FCA). The boat was originally built as an oil-rig supply vessel in the early 1970s. In the late 1980s, it was converted to a factory fishing trawler. *Photograph by Ed Cook*

To catch fish, the trawlers lower a huge net off the stern and drag it across the ocean floor for miles at a time. Then the fishermen behead, gut, and freeze the fish in a factory one level below the ship's trawl deck. *Photograph by Ed Cook*

The view from the deck of the *Alaska Ranger* in February 2008, as the ship fished in ice near the Pribilof Islands. *Photograph by David Hull*

The crew of the Coast Guard's 6007 Jayhawk helicopter was the first to reach the sinking site on the morning of March 23, 2008. *(Left to right:)* Aircraft Commander Brian McLaughlin, pilot Steve Bonn, rescue swimmer O'Brien Starr-Hollow, and flight mechanic Rob DeBolt. *Photograph by Byron Cross/USCG*

The Coast Guard's Hercules C-130 number 1705 took off from Elmendorf Air Force Base in Anchorage within half an hour of getting the report of the sinking ship. It would take the search plane several hours to reach the disaster site. *Photograph by Evan Isenstein-Brand*

The Coast Guard's 378-foot cutter *Munro* raced toward the *Alaska Ranger* on its turbine engines at close to 30 knots.
Photograph by Kurt Fredrickson/USCG

Like all Coast Guard air crew members operating over cold water, Jayhawk pilot Brian McLaughlin wore an orange drysuit with a snug rubber seal at the neck and a lightweight personal floatation device (PFD), which inflates with the tug of a drawstring. The pilot's helmet is equipped with night-vision goggles and an audio system that allows the rescuers to communicate over the deafening thud of the helicopter rotors. *Photograph by Henry Leutwyler*

Helicopter crews can use a metal rescue basket to pull up victims from the water. Though the basket is rated to 600 pounds, it is extremely rare to lift more than one person at a time. *Photograph by Henry Leutwyler*

Rescue swimmer O'Brien Starr-Hollow saved the lives of more than a dozen of the *Alaska Ranger*'s crew members. He wore a neoprene hoodie under his reflective helmet, as well as a mask, snorkel, and fins. The yellow arms and reflective tape on his drysuit allowed Starr-Hollow to communicate in the dark with his flight mechanic in the helicopter above. *Photograph by Henry Leutwyler*

Captain Eric Peter Jacobsen had been working for the FCA since the mid-1980s, but he had only been the captain of the *Alaska Ranger* for a few weeks. *Photograph courtesy of Karen Jacobsen*

Like all FCA ships, the *Alaska Ranger* had a Japanese fish master, Satoshi Konno, whose official job was to direct the trawler to the best fishing grounds. *Photograph courtesy of Richard Canty*

Federal fisheries observer Gwen Rains was the only woman among the forty-seven people aboard the *Alaska Ranger* on March 23, 2008. She boarded the ship for the first time just four days before the disaster occurred. *Photograph courtesy of Gwen Rains*

Alaska Ranger processor Julio Morales (*far left*) and his younger cousin Byron Carrillo (*far right*) were raised together by their grandmother in Guatemala after their mothers both left for the United States. Neither man had ever worked on a fishing boat before they joined the crew of the *Alaska Ranger* in March 2008. *Photograph courtesy of Julio Morales*

David Silveira normally sailed as the captain of one of the FCA's long-liners, the *Alaska Pioneer*. He'd agreed to step in—temporarily—as the *Alaska Ranger*'s first mate. *Photograph courtesy of Celeste Silveira*

Chief Engineer Dan Cook had been working on fishing boats since he was fourteen years old, but had only been on the *Alaska Ranger* for a couple of months. *Photograph courtesy of Ed and Cindy Cook*

Bering Sea fishing boats are required by law to carry a neoprene survival suit for each man on board. If the "Gumby" suit is a good fit and worn properly, very little water should leak inside.
Photograph by Henry Leutwyler

Just hours before being called to rescue the men on the *Alaska Ranger,* several of the air crew members aboard the Coast Guard's cutter *Munro* participated in an eating contest. (*Left:*) Rescue swimmer Abram Heller. (*Center:*) Pilot Greg Gedemer gulps down a spoonful of cold baked beans. *Photograph by Greg Beck/USCG*

As part of their training, federal fisheries observers practice swimming in their survival suits in a pool in Anchorage. They are taught how to link together in the water and form a "pinwheel" to signal to a potential search plane overhead. *Photograph courtesy of the author*

The Coast Guard's 6566 Dolphin helicopter was deployed onboard the *Munro* during the ship's March 2008 Bering Sea patrol. Here, the forty-five-foot aircraft lands on the flight deck on the morning of the rescue. *Photograph courtesy of cutter* Munro/*USCG*

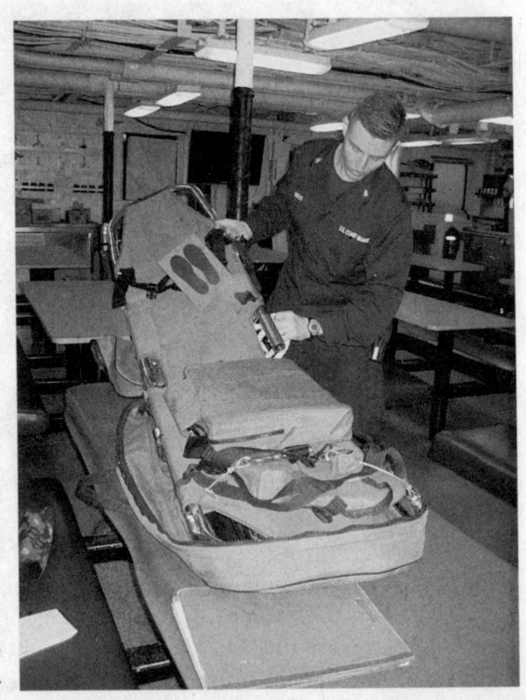

The ship's corpsman, "Doc" Chuck Weiss, displays the litter used to carry some of the *Alaska Ranger*'s crew members down onto the *Munro*'s mess deck, which became a temporary infirmary. *Photograph courtesy of the author*

The crew of the 6566 Dolphin (*left to right*): pilot Greg Gedemer, Aircraft Commander TJ Schmitz, flight mechanic Al Musgrave, and rescue swimmer Abram Heller *(being held). Photograph courtesy of USCG*

Julio Morales is helped across *Munro*'s deck after being lowered in the rescue basket from the 60 Jayhawk. The larger helicopter is too big to land on the ship's deck. *Photograph courtesy of cutter* Munro*/USCG*

The 60 Jayhawk used a method known as helicopter in-flight refueling (HIFR) to gas up from the cutter *Munro*. *Photograph courtesy of cutter* Munro*/USCG*

A Coast Guard rescue swimmer examines an empty life raft from the *Alaska Ranger* on the afternoon of March 23, 2008. The swimmer punctured and sank the raft so that it would not offer false hope to fellow searchers. *Photograph by Dan Lytle/USCG*

This photograph was taken from inside the 60 Jayhawk as it hovered beside the *Munro* during HIFR. *Photograph by Byron Cross/USCG*

The *Munro*'s commander, Captain Craig Lloyd, at the Coast Guard dock in Dutch Harbor. *Photograph by Charles Homans/AP*

The FCA trawler *Alaska Warrior* at dock in Dutch Harbor.
Photograph by Ed Cook

Members of the Marine Board of Investigation, which was convened to determine the cause of the sinking of the *Alaska Ranger,* were led on a tour of the *Warrior* a few days after the disaster. Pictured here are Coast Guard Captain Mike Rand *(left)* and National Transportation Safety Board (NTSB) Investigator Liam LaRue *(second from right).*
Photograph by Sara Francis/USCG

The *Alaska Ranger*'s used survival suits were laid out for examination by the Marine Board members. *Photograph by Sara Francis/USCG*

The investigators examine the *Warrior's* trawl deck, which is crowded with fishing nets, lines, and buoys. *Photograph by Sara Francis/USCG*

Captain Pete Jacobsen's daughter, Karen, and the *Alaska Ranger*'s long-time cook, Eric Haynes, at the Seattle Fishermen's Memorial on Easter morning, 2009. *Photograph courtesy of the author*

Alaska Warrior Chief Engineer Ed Cook photographed one of the *Warrior*'s water-tight doors tied open at sea on April 29, 2008—just over a month after the *Alaska Ranger* tragedy. A plaque bolted to the door reads "Keep Closed at Sea." *Photograph by Ed Cook*

appear dead: unconscious, with bluish-gray skin, dilated pupils, no pulse, and no heartbeat. Still, according to Steinman and other cold-water survival experts, rescuers should attempt CPR. It is possible for some cold-water victims to be revived, without brain damage, even an hour after they've lost consciousness.

Most deaths that occur within the first half hour of cold-water immersion are not true hypothermia. Rather, they are due to panic and the subsequent problems associated with people's inability to control their breathing or their thinking. Even in the coldest waters, hypothermia usually takes at least half an hour to kill. Nonetheless the majority of drowning deaths in cold water are in fact a consequence of hypothermia. A victim goes through the states of hypothermia to the point of losing consciousness, then drowns because he can't keep his head out of the water.

Regardless of survival equipment, some people are inherently more apt to survive than others. Both anecdotal and laboratory evidence supports the thesis that increased body weight boosts the likelihood of survival. It makes intuitive sense that, as Steinman writes, "smaller adults generally cool faster than larger adults and tall, lanky individuals cool faster than short, stout individuals. . . . Fat is a very efficient insulator against heat loss."

Some of the most-repeated "miracle" stories of long-term survival in Alaskan waters involve individuals who were heavier than average, and a 1985 study found that the effect is, in fact, significant. Very thin subjects in the tenth percentile of skinfold thickness (a measure of subcutaneous fat) wearing light clothing in 41°F water were found to cool at *nine* times the rate of their portliest peers (those in the ninetieth percentile of skinfold thickness).

* * *

EVERY TIME HE ROSE UP TO THE TOP of a wave, Evan Holmes could see other lights spread out in the ocean. He could hear people yelling. Then he would sink down into another trough and *whoosh,* the yells would disappear along with the lights. The crashing waves were so loud, they blocked it all out. Evan was on top of a swell when he saw his ship for the last time. The *Alaska Ranger* was stern-down in the sea, the bow pointed straight toward the sky. *Whoosh,* Evan plunged down with a wave. The next time he rose up, the *Ranger* was gone.

Evan had been in the water for maybe an hour when he saw someone floating toward him and P. Ton, who was still holding on to Evan's legs. When the guy got closer, Evan could see that it was Kenny Smith.

"Kenny! Come on, let's make a chain!" Evan yelled over the crash of the waves.

The factory manager knew that the other guys hadn't been to the safety course in Seattle like he had. Now, the lessons he learned there were paying off. Evan showed Kenny how to link together, with one man's legs wrapped around the other's waist.

Both Kenny and P. Ton were smaller than Evan. They both seemed colder and in worse spirits. "We're not gonna make it," Kenny kept saying.

"Keep calm!" Evan yelled. "If you don't shut up, I'm gonna give you overtime!"

After a while, the men saw someone else floating toward them in the water. Kenny grabbed the man's suit.

"Evan, man. Evan! This guy is dead. He's dead, he's gone," Kenny screamed.

Evan touched the body for a second. It was lifeless. "Oh my God," he said. "Shit."

Kenny couldn't tell who the guy was. Someone small. Evan let go, and the body floated quickly away.

* * *

IT WAS FREEZING INSIDE THE JAYHAWK HELICOPTER. In the more than an hour since the crew had taken off from St. Paul, they hadn't been able to turn on the heat inside the aircraft. The helo's heating system and its deicing system both use the same hot air off the engines. The night was cold enough that all the heat had to be used to keep the engines on anti-ice.

Under normal conditions, the Jayhawk burns about 1,000 pounds of fuel an hour (the equivalent of 150 gallons) and the deicer increases that to 1,100, 1,200, even 1,300 pounds depending on weather conditions. To run the heat and the deicer at the same time, they would have had to bring in another generator, which would burn fuel even faster. Heat was a luxury they couldn't afford.

Luckily, the men were warmly dressed. All four of them wore bright orange, fire-retardant dry suits, the required uniform for Coast Guard helicopter crews operating over waters colder than 70°F. With snug rubber seals at the neck and wrists, the suits are designed to keep rescuers dry, even if they end up completely submerged. The material, though, breathes just enough to keep the wearer from feeling swampy after the inevitable hours in the air. Rescue swimmer O'Brien Starr-Hollow wore an extra layer of synthetic long johns beneath his dry suit. He was the only one of the four men, after all, who planned to be getting in the water.

As they neared the scene, Starr-Hollow disconnected his flight helmet from the Jayhawk's Internal Communication System (ICS) and started to gear up. He checked the seals on his wrists and neck, making sure there wasn't any fabric breaking the secure bond between suit and skin. He pulled on his 7mm-thick neoprene hood and a neon yellow helmet, similar to those worn by white water kayakers. He checked that the rest of his gear was ready to go. Right before entering the water, he'd replace his fire-retardant flight gloves with neoprene hyper-stretch wet gloves and pull his fins on over his work boots, which he wore over the dry suit's built-in booties. Last would be his mask, with a mini dive light attached, and snorkel.

During flight, the rescue swimmer was responsible for backing up the pilots by keeping a constant eye on the radar and fuel burn, and for running the radios from the back of the plane. There was a guard established over the HF radio. Every fifteen minutes, Starr-Hollow would check in with COMMSTA Kodiak, which would record the helo's GPS position, direction, and speed. If the aircraft went down, COMMSTA would have a reasonably good idea of where to search for survivors. Starr-Hollow was also responsible for backing up the pilots on altitude—if they started descending, he'd be calling out the altitude over the ICS. It was a life-or-death job in whiteout conditions with twenty-foot breaking waves.

THE RESCUERS ALL KNEW THE STORIES of aircraft that had been taken out by a rogue wave. It had happened in the Bering Sea just a few years before, during a rescue attempt off a grounded Malaysian cargo ship, the *Selendang Ayu*. A Coast Guard aircrew and a helicopter full of survivors was batted out of the sky by a monster swell, unseen until it was too late. The flight crew had all been wearing helmets and dry suits. Most important, they'd been trained to escape a submerged capsule.

It's part of the standard training for every Coast Guard airman. The Navy has eight facilities around the country equipped with "helo-dunkers," mock helicopter pods suspended on a crane above the deep end of a pool. During their initial training, and every six years afterward, each member of a Coast Guard flight crew is sent to one of the Navy facilities to practice escaping a downed chopper. The crane lowers the dunker to the surface of the pool, and the rescuers, dressed in full flight gear, climb inside and buckle themselves in. Then, the dunker is raised above the water, dropped—and spun. A helicopter's rotors and gears make the machine top-heavy; in a crash into water, the aircraft is likely to flip upside down. The rescuers are trained to open the aircraft doors and push out the windows *before* the helo hits water, if possible. Once they're under, they learn to unbuckle themselves from an inverted position and pull out a tiny scuba tank known as the HEED, for helicopter emergency egress device (all members of the crew fly with the soda-can-size air bottle in a pocket).

The crew members' most crucial lesson is to keep constant contact with a reference point on the inside of the submerged aircraft, and to travel hand-over-hand to get themselves out a window or door and up to the surface. Let go and all sense of up and down is lost, Coasties are taught. After mastering escape with the air bottle, the crew learns to escape while wearing blacked-out goggles. Finally, they do it with blacked-out goggles and *without* supplemental air. They practice holding their breath for longer than they thought possible. They learn not to panic when the worst happens. And when it does, they sometimes survive.

In that December 2004 *Selendang Ayu* disaster, the training worked: Each member of the helicopter crew got out safely. Six of the seven sailors they'd just plucked off the seven-hundred-foot freighter were killed in the crash.

EVAN HOLMES HAD BEEN SHAKEN UP by the dead body. He was cold, and he had a little water in his suit. When he lifted his arm, he could feel a trickle of icy water run down toward his chest. Evan was worried about hypothermia setting in. Crap, we've been floating around for quite a while, he thought. He couldn't stop shivering. The other guys seemed just as bad—maybe worse.

Kenny and P. Ton had been real quiet ever since they saw the body. Evan tried to think of a song to sing. He should try to keep them all occupied with something other than the fact that they didn't know if they were going to make it out of the Bering Sea. But for the life of him, he could *not* think of single song.

"Hey, Holmes, I'm not gonna make it," Kenny was saying.

"Yeah, you are," Evan told him. "You are."

Evan wasn't a religious guy. To him, it seemed a little selfish to start praying just for his own life at a time like this. He looked up into the dark sky. Here's the deal, God, Evan bartered. Give us one more sunrise. We want to see the sun one more time. If I'm going to be floating around in this ocean like a Popsicle, I want to see the sun rise just once more.

Chapter Eight

Swimmer in the Water

Jayhawk pilots Brian McLaughlin and Steve Bonn scanned the waves. It was almost 5:00 A.M., but in Alaska in wintertime, 5:00 A.M. still looks like the middle of the night. Attached to their flight helmets, the men wore night vision goggles, heavy metal optics that gave the entire ocean the neon green glow of an old-school video game.

Finally, the helicopter broke out from a snow squall, and there it was—a light. Then two, three . . . five. The men saw what looked like a poorly lit runway, a ragged string of strobes flashing on and off over a mile-long stretch of ocean. They scanned the seas for a ship. But there was no sign of the *Alaska Ranger.*

The scene was unlike anything the four Coast Guard rescuers had ever faced in the past. McLaughlin stared down at the ocean one hundred feet below. To his left, to his right—everywhere he looked he saw more blinking strobes. There were at least two dozen individual lights spread about in the waves.

Oh my God, he thought. Where do we begin?

The men knew that the *Munro* was making its way toward the disaster site, racing on its turbine engines at close to 30 knots. Still, the ship was hours away. And given the sea conditions, McLaughlin thought the *Munro* most likely wouldn't be able to launch its rescue helicopter.

His aircraft was it, the only hope for these people—at least for now. They just had to choose a spot and start getting people out of the water.

ERIC HAYNES HAD BEEN INSIDE THE LIFE RAFT for what seemed like an hour when he heard the rotors. The Coast Guard helicopter was overhead. Out the open flap of the life raft's door, Eric noticed the full moon above. Thank God for that, Eric thought. It might help those guys in the water, and the Coasties here to save them.

The ten men in Eric's raft had two handheld radios. Boatswain Chris Cossich had grabbed one of them right before he abandoned ship. Ever since he got situated in the raft, Chris had been repeating "Mayday, Mayday," and waiting for a response. Now, he heard the Coast Guard calling over Channel 16.

"We see you," McLaughlin told Chris. "Are you okay?"

"There's ten of us," Chris told the chopper pilot. "Some guys are cold, but we're all right."

"We're going to start with the people in the water," the Coastie said. "Bail out as much water as you can, stay close to each other, and try to stay warm. We'll come back for you."

* * *

Bonn pulled the aircraft over the first light the Jayhawk reached. It was one guy, alone, but alive. The whole crew could see him waving. The pilot flew a lap over the scene. There were people everywhere. Everyone they could see was in a survival suit, and no one looked obviously worse off than anyone else. Not that that was an easy judgment to make from the air.

The aircrew had the dewatering pump with them; many times in the past, a pump had been enough to solve a crisis at sea. But there was nothing left to save. DeBolt and Starr-Hollow pitched the pump out the aircraft door to make more room in the cabin. They had also brought along one of the Coast Guard's mass casualty life rafts, which was made to hold twenty people, the same number as the *Alaska Ranger*'s. There was a long line attached; a sharp tug should activate a CO2 cylinder to inflate the raft. It was best to hold the line, kick the raft out the aircraft door, then yank the rope when the raft hit the water. DeBolt and Starr-Hollow punched the life raft out the door, but the line had a knot, and ripped out of Starr-Hollow's hands before the raft hit the surface. They'd chosen to drop it in a spot where an inflated raft might float downwind to some of the survivors. But now they couldn't see it; they had no idea if it had inflated or not. The raft was gone.

Bonn pointed the helicopter back toward the first guy they'd seen. He was the farthest downwind; he'd probably been in the water the longest, the pilots guessed. They'd get him first.

Until this week at St. Paul, the men in the Jayhawk had never before flown together as a four-man team. Most had worked with each of the others at some point in the past, though. Just a couple months before, McLaughlin and Starr-Hollow were both on duty in Kodiak when they got a call from a fishing boat that was taking on water with four people and a dog on board. When they got to the site in Shelikof Strait, on the far side of Kodiak Island from the air station, everyone had already abandoned ship into a life raft. Starr-Hollow descended into the water and brought them all up—the little dog, too. By the time everyone was safe inside the helicopter, the fishing boat had fully disappeared beneath the waves.

It had been daytime and the conditions were relatively calm, but it was still a big case. It's rare to airlift so many people at once. McLaughlin had been grateful to have Starr-Hollow as his crew's swimmer that day. All the rescue swimmers in Kodiak were in phenomenal shape, and obviously game for the job. If anything, McLaughlin had seen swimmers get pissed off when they flew on a case and then weren't used. But even among an impressive crowd, Starr-Hollow's intense calm and focus stood out. It was hard to imagine him running out of energy. If you were going to be out in the stink, O'Brien Starr-Hollow was the kind of guy you wanted to have out there with you.

The thirty-three-year-old rescue swimmer had requested Kodiak as his first assignment out of A School. He'd been there since 2003. He wasn't the typical new swimmer. Most are in their early twenties, some as young as nineteen. Starr-Hollow didn't join the Coast Guard until he was twenty-seven. He grew up in Helena, Montana, and was an alpine ski racer and soccer player in high school. After graduating in 1992, he enrolled at Oregon State University in Corvallis, where he rowed crew for all four years. He loved the total commitment and teamwork of being an oarsman. Starr-Hollow's father had been a Navy SEAL, and early on, Starr-Hollow signed up for Navy ROTC. But the training conflicted with rowing, so he gave it up.

After college, Starr-Hollow worked as a ski coach and then started a geology grad program in Missoula. By the summer of

2000, he was in eastern Montana, working on water quality and riparian ecology research. There were a lot of wildfires that year and a lot of firemen around. It was backbreaking work they did, and Starr-Hollow found himself wishing he was doing something more like it. Though he was spending his days outside, there wasn't much physical labor in his scientific work. He envied the camaraderie he saw among the firefighters and the pure physicality of their job. It was something he missed from his rowing days.

"The grass is always greener over the septic tank," Starr-Hollow's mom said when he told her he was thinking of joining the Coast Guard and becoming a rescue swimmer. It was one of her favorite expressions. Starr-Hollow figured he could always go back to science—he planned to, in fact. But at that point in his life he needed something different. In August 2001, Starr-Hollow showed up at boot camp in Cape May, New Jersey.

At twenty-seven, he was one of the oldest in his class. He was instructed in the basics of good seamanship, learned to march in formation—and was taught how to fold his socks. The method was to roll them up tightly and then flip the last little lip over so the ball had a smiley face in front. When socks were lined up the right way in a locker, there'd be a row of smiley faces. Starr-Hollow was given a map of how to organize his clothes. On the ships, in tight quarters, it was important to keep things neat and organized.

There wasn't as much physical training in boot camp as he'd expected, although the discipline was intense. Starr-Hollow felt like he never got enough sleep or enough to eat. They were kept busy from six in the morning until ten at night. It was a trudge, but Starr-Hollow tried to make the best of it. He hadn't played the saxophone since his sophomore year in high school, but he picked it up again to join the Coast Guard band. Some evenings they performed at the bandstand in downtown Cape May—marching songs for the town residents. He started going to church, mostly to have some time away from the drill sergeants.

Starr-Hollow had heard of guys getting stuck in port for a year or just patrolling tiny, monotonous regions aboard the Coast Guard's 378-foot cutters. "The white needle of death," guys called those boats. Instead, Starr-Hollow requested the ice-breakers; he figured they would definitely travel. He was right. His first assignment out of boot camp was on the 399-foot *Polar Star.* In the next six months he saw Hawaii; Sydney, Australia; and Hobart, Tasmania. The ship spent a month breaking a route into McMurdo Station in Antarctica, the main support facility for American researchers at the bottom of the world. On the way home, they made port calls in Chile, Peru, Mexico, and San Diego before returning back to Washington State.

A month later Starr-Hollow started his airman program in Port Angeles, essentially an internship in the job of Aviation Survival Technician (AST), the official title that encompasses the job of rescue swimmer. Flying as the deployable aircrewman was only a tiny part of an AST's responsibility, Starr-Hollow quickly learned. Ninety-eight percent of the job was about equipment and preparation. The ASTs were responsible for keeping all the helicopters' emergency gear in serviceable condition and for preparing themselves for the 2 percent of the job that mattered most—saving lives. Starr-Hollow trained with the swimmers in Port Angeles for four months before heading to sixteen weeks of A School in Elizabeth City, North Carolina.

It's an extremely competitive program. Almost half of the students who start rescue swimmer A School fail out. The physical standards are intense—and so are the psychological ones. For Starr-Hollow, the hardest part was the "bullpen." The instructors are all in the deep end of the pool, while the students stand blindfolded at the shallow end. One by one, the students are called down to the deep water. There, they are circled by the instructors, who each take a turn wrestling each student down to the bottom of the pool. The student has to break free of each instructor, and bring him back up to the surface in a controlled carry. The goal is to prepare the would-be swimmer to handle a panicked survivor, and the instructors play the part, kicking and swinging punches at the young students. The students who can handle it, who don't freak out, will likely be able to remain calm in even the most chaotic real-life rescue.

Starr-Hollow graduated from A School and was sent straight to the shop in Kodiak. The work was satisfying—and so was the lifestyle. In Alaska, he could backcountry ski for a good part of the year, surf when the swells grew large in the winter, and get in a little mountain biking when the trails weren't a torrent of mud. He met his wife there. She had a young daughter, and they soon had another little girl together. In the winter and spring, he would wake up early before the girls were out of bed and ski Pyramid Mountain. In two and a half hours he could hike up the peak, ski down, and drive back home, "gassed" but happy.

RYAN FELT LIKE HE'D BEEN in the water for days. But it was still dark; it couldn't have been more than a few hours. He was still thinking about unzipping his suit. When should he do it, how long should he wait? Then he saw a light way off on the horizon. A ship! he thought. The *Warrior,* maybe. He knew how long it took between when you spotted another ship in the distance and when you actually passed it side to side. He figured the boat was more than an hour away. But the light was growing closer quickly. No more than thirty seconds after seeing it, Ryan heard the rotors.

The chopper seemed to home in right on him. It approached like a missile, and stopped short just above him, maybe one hundred feet into the sky. A giant spotlight shone down. Ryan waved his arms. For a few seconds the orange machine hovered above him. Then it turned and flew away.

What the hell, Ryan thought, I know they saw me.

He kept his eyes on the helicopter as it made a giant lap over the ocean. Then thankfully, miraculously, it circled back and settled over him. The door swung open.

He was going to be saved.

As Bonn came into a hover over Ryan Shuck, Starr-Hollow clipped his Triton harness into a talon hook on the end of a metal cable that ran into a hoist hard-mounted to the outside of the Jayhawk. At DeBolt's signal, Starr-Hollow slid forward to sit with his legs dangling over the edge of the open aircraft door and unclipped from his gunner's belt.

"Ready for direct deployment of rescue swimmer to survivor," flight mechanic DeBolt announced through the ICS. "Swimmer is at the door." And then, using a mechanical control just inside the aircraft door, DeBolt retracted the hoist cable, drawing Starr-Hollow smoothly up and out of the aircraft.

"Swimmer is outside of the cabin," DeBolt reported. "Swimmer going down."

From the pilot's seat, Bonn couldn't see much of what was going on behind him in the cabin, or in the waves beneath the aircraft. Through the ICS, a Coast Guard flight mechanic paints a verbal picture of the scene below, constantly updating the pilots with the information necessary to keep the helo positioned safely above the swimmer and victims in the water. Standard procedure during a hoist operation is for the flying pilot to turn off his or her radio—about 7 percent of Coast Guard helicopter pilots are women—and to concentrate only on the flight operations and the instructions of the flight mech. Meanwhile, the copilot handles all communication with people outside the helicopter. As soon as the decision is made to lower a swimmer, the flight mechanic is running the show, feeding the rest of the crew a constant stream of commands.

The conversation is scripted, the language drilled into all aircrew members from the first days of their training. From the safety checklists that the crew collectively runs through every time the swimmer leaves the cabin, to the "conning"—or positioning—commands that keep the helo safely placed over breaking swells, the crews are speaking a custom-made language built on succinct, declarative sentences. In the middle of the night, in the Bering Sea, for even one member of a four-man helicopter crew to be confused about what's happening is to put the entire crew in danger. Precision. Clarity. Those are the attributes, each Coast Guard rescuer had been taught, that allow even the most complicated or harrowing rescue to proceed smoothly and calmly.

Now that the swimmer was out the door, McLaughlin would be the second eyes on the helo's altitude. He continually scanned the gauges that covered the panel in front of him, and called out the size and frequency of the incoming swells to help DeBolt manage the hoist. The strong wind was working in the rescuers' favor. Often the Jayhawk's 100 mph rotor wash overwhelms people in the water, but with the gusts off the nose blowing most of the rotor wash behind the helicopter, the rescuers were able to fly closer to the survivor than usual.

DeBolt was kneeling at the open cabin door, attached to the

roof by a canvas gunner's belt that would hold him to the helo even if he tumbled out the opening. It was DeBolt's job to make sure the hoist cable didn't become tangled around itself, in the aircraft's landing gear, or around a person in the water. As he lowered Starr-Hollow toward the surface, DeBolt leaned out into the darkness, eyes glued to his swimmer's neon yellow helmet.

"Swimmer going down. Swimmer halfway down," he continued through the ICS.

From the waves, Ryan Shuck saw the bright light of the helicopter fifty feet above, and the outline of the rescue swimmer falling slowly toward the surface. Ryan started swimming as fast as he could toward the rescuer.

As Starr-Hollow sank toward the waves on the thin metal line, he could see the fisherman trying to swim for him. With his harness clipped into the talon hook at the end of the hoist line, the Coast Guard rescuer had both his hands free for swimming or to grab on to a survivor. But he wasn't close enough. Starr-Hollow seemed to be bobbing up and down, twenty feet at a time, as the waves swelled and retreated beneath him. As the aircrew struggled to keep the helo stable, the rescue swimmer found himself skimming forty or fifty feet horizontally over the water. Meanwhile, Ryan, too, was pushed all over by the waves and was fighting to move toward Starr-Hollow.

Finally, DeBolt placed the rescue swimmer just feet from Ryan, who reached out toward Starr-Hollow as the swimmer called to him.

"Swimmer in the water," the flight mech announced through the ICS as Starr-Hollow hit the waves.

Ryan watched as the rescuer hit the ocean waist-deep, and was carried right to him by the wind.

"Stop swimming!" Starr-Hollow yelled.

DeBolt fed out more cable and directed Bonn to back the aircraft away from the men in the water. Starr-Hollow would remain attached to the hoist; feeding out the extra cable would give him some more room to maneuver, while backing away would help ensure that the extra line didn't get tangled in itself, or worse, around their swimmer or survivor. If the line suddenly became taut—from aircraft movement or from a large wave dropping out from under the men in the water—Starr-Hollow and the fisherman could be jerked violently out of the waves.

"Swimmer is at the survivor," DeBolt announced as, forty feet below, Starr-Hollow grabbed on to Ryan's arm.

Ryan could feel the swimmer's strength instantly. He began to relax.

"How're you doing?" Starr-Hollow yelled over the thud of the rotors above.

"Okay," Ryan answered. "Okay."

Starr-Hollow told Ryan his name, and asked if Ryan could keep his arms still, as tight as he could against his sides.

"Yes," Ryan nodded, "I can do whatever you want me to do."

Starr-Hollow had a simple harness, called a rescue strop, slung over his shoulder. It took him ten seconds to cinch it over Ryan's chest and clip the tightened harness into the talon hook on the end of the hoist line. Then he gave DeBolt the thumbs-up, the signal to hoist them out of the water. The hook would carry the weight of both men as they were pulled up, Starr-Hollow holding Ryan's head and torso steady between his own legs.

The harness felt tight around Ryan's chest, but the ride was quick. Within fifteen seconds, he was crawling toward the back of the Jayhawk and Bonn was flying toward the next closest light, about one hundred yards away. They came into a hover, and again Starr-Hollow went down on the cable with the quick strop while Bonn maneuvered the helicopter in response to DeBolt, who was reporting the swimmer's every movement, and McLaughlin, who was calling out the size and frequency of the incoming waves.

"I knew you guys would make it," the second fisherman said to Starr-Hollow when the swimmer reached him in the breaking seas. It was hard to have much of a conversation above the noise of the helicopter. But again, Starr-Hollow secured the fisherman in the strop and explained to him that he had to keep his arms down, straight at his sides to prevent the simple device from slipping up and over his head. Then he gave DeBolt the thumbs-up.

IN THE FOUR YEARS HE'D BEEN in Kodiak, Starr-Hollow had twice been sent for a week of advanced training on the coast of Oregon. The Advanced Helicopter Rescue School, known as AHRS (pronounced "Arse") is held eight times a year, in late fall and early winter. The course is timed to take advantage of the huge swells that form where the Columbia River meets the Pacific Ocean on the border of Oregon and Washington. In five mornings of instruction, the students—eight swimmers, four mechanics, and four pilots—are schooled in advanced rescue techniques in a classroom in Astoria, Oregon. In the afternoons they fly and train in the Middle Grounds of the Columbia River Bar and on the cliffs near Cape Disappointment, Washington, where Lewis and Clark first saw the Pacific in 1805. Tourists and locals alike often gather on the beach to watch the helicopter teams pluck dummies from the crumbling cliffs.

All four men in the Jayhawk had been to the Astoria school, where they practiced conning commands in the massive offshore swells. Back in Kodiak, they flew training missions all the time. Every day of the week, at least a couple of the air station's helos would be up, practicing for every possible scenario. They would simulate engine failures, fires in the cockpit, and anything else that could potentially go wrong with the aircraft. They practiced dropping their swimmer to small boats and into the water and hoisting him up again in different situations. They had standards to meet. Every six months, each crew member had to complete a set number of trainers to maintain their qualifications to fly.

The training at the Advanced Helicopter Rescue School, though, was a level beyond the norm. Even though Alaska has the most extreme flying in the country, the crews have Coast Guard–wide weather limits for training. At AHRS, these limits are lifted a bit and the crews have the chance to practice in conditions they rarely—if ever—encounter at their home stations.

Now that training was paying off. Normally, Bonn and McLaughlin would never drop their swimmer into seas as violent as these just for practice. But at AHRS, the waves had been pretty similar. They'd had the opportunity to see exactly how people in the water—people tethered to their aircraft with a potentially deadly cable—reacted to enormous seas. And they'd had a chance to practice the advanced communication skills needed to simultaneously manage a complicated hoist and difficult weather conditions.

After a while, the training at the Columbia River Bar had begun to feel like clockwork. Tonight, they would get into the same rhythm. Each member of the crew knew what to do. Now, methodically, step by step, they'd do it.

WITH TWO SURVIVORS SAFELY INSIDE THE HELICOPTER, Bonn piloted the helo over to a group of people who had linked arms in the water. The quick strop system is the fastest way to get a single, conscious, cooperative victim out of the ocean. Now the crew decided to switch from that harness to their metal rescue basket. The basket is generally considered a safer hoisting option for anyone having trouble breathing. The quick strop pulls tightly around a survivor's chest, but there is no strap in the basket. The victim simply sits inside the high-walled box until it's brought inside the helicopter. The basket is also faster for lifting multiple victims because the rescue swimmer stays in the water to prep the next person while the first person is lifted.

DeBolt lowered Starr-Hollow into the water on the hoist line, a few yards from the survivors. The swimmer unhooked his harness from the cable, and swam toward the group of fishermen. Meanwhile, DeBolt drew the cable back up to the aircraft door, grabbed the hook, and attached it to the top of the rescue basket's metal bales.

Starr-Hollow quickly swam up and down the line of men. There were six of them, locked elbow to elbow. They were getting twisted and turned and washed over by the waves. The swimmer's first task was to determine who was the worst off—who would go first.

The swimmer could tell that the fishermen had little mobility in the thick Gumby suits. Watching them struggle in the water was like watching someone trying to run a race in a sleeping bag, Starr-Hollow thought. Compared to the fishermen, he was competing in jeans and a T-shirt. The buoyancy of Starr-Hollow's dry suit was almost neutral, making it relatively easy for the rescuer to swim among the swells. A Gumby suit, on the other hand, has a buoyancy of almost 40 pounds. Even a very strong man would have great difficulty diving down below a breaking wave while wearing one.

Starr-Hollow peeled the first survivor from the group, swam

him over a few yards away from the other fishermen, and signaled for DeBolt to drop the basket. The mesh metal box has red floats on either end—each printed with the words "stay seated"—that prevent it from sinking. The basket is rated to hold up to 600 pounds, though in practice, it's extremely rare to lift more than one person at a time. There'd been cases in which women and children, and very slight men dressed in street clothes, had been brought up two at a time. The typical fisherman? Even without a Gumby, he was going up alone.

As Starr-Hollow helped the first man into the basket, he held the excess cable away from the compartment to avoid tangling. When DeBolt began raising the basket, Starr-Hollow hung his weight onto the bottom. He used his body like an anchor until the last possible moment, preventing the basket, and its passenger, from swinging too wildly above the waves.

Forty feet above, DeBolt maneuvered the first man into the cabin as Starr-Hollow swam back to the group of fishermen. The bottom of the basket can be raised just above the edge of the aircraft floor. The flight mechanic grabs the basket, pulls it into the cabin, and tips it on its side to allow the passenger to climb out. The first couple of guys from the group of six easily crawled out of the basket and huddled into the growing mass of fishermen against the back wall of the Jayhawk. As each man emerged, he was embraced by those who'd come before—most of them still barely recognizable in their hooded suits. By the time the third man was up, each new arrival brought a round of cheering and high-fives.

Then DeBolt pulled a smaller-framed man into the cabin. The flight mech got the basket into the helicopter, but the fisherman inside was paralyzed with fear. *This guy is obviously in shock*, DeBolt thought. The fisherman wouldn't budge from the basket. He looked scared to death. Over the deafening thud of the rotors, DeBolt yelled for the men huddled to the side of the cabin to help him get their friend out of the basket. Several men grabbed onto the fisherman's fingers, attempting to pry open his grip on the metal rails. It was Joey Galbreath, a processor from Washington State. When the others called his name he just stared straight ahead like he was in a trance and wouldn't let go. It took several minutes, even with DeBolt and the fishermen working together, to pull Joey out of the basket and get it back out the door toward the next man in the water.

EVAN HOLMES WAS CONFIDENT that the helicopter had seen them. It had been directly overhead; the spotlight was right on them. He'd waved his arms.

"They know we're here," he told P. Ton and Kenny. "They see us. They're coming back. We just have to hang in there."

Evan was in the rear of the three-man chain. He was the only one who had his arms free, and he kept trying to steer the other guys so that the waves would hit them all in the back instead of the face. Each time they were pummeled from the front, water leaked down into their suits. It was so dark, though, that it was hard to see where the water was coming from. It was only every once in a while that a little sliver of moon poked through the clouds. When they rose up on the top of a swell, Evan could see the helicopter in the distance, hovering. Then they'd plunge back down, and the chopper would fall out of sight.

Finally, the light came back, and settled overhead. "No shit," Evan said. "Thank you." He watched the rescue swimmer drop down from the open door. "U.S. Coast Guard!" Starr-Hollow yelled as he reached the men in the waves.

Evan pointed to Kenny. The young processor was nearly

unconscious. A few minutes before, he'd started drifting away from the other men. He couldn't hold on to P. Ton's legs anymore. Evan feared that Kenny was done.

"Hey, man," Evan yelled at Starr-Hollow. He motioned toward Kenny. "This guy, he ain't doing too well. Get him first."

Starr-Hollow could tell on his own that Kenny was the worst off of the three men. He loaded him into the basket, gave DeBolt the thumbs-up, and then swam back toward Evan and P. Ton.

"How are you doing?" Starr-Hollow yelled.

"I'm okay," Evan answered. "Well, I'm not okay. Look, I'm doing okay right this second. I don't know how much longer, though."

Evan had seen how long the helo had been out there picking people up, and he'd seen some movies. He knew P. Ton should go up next. "Shit, man, do you have room in that helicopter for me?" he yelled to the swimmer.

Starr-Hollow grabbed onto Evan and looked him right in the eyes. "I tell you what," he said. "I've got room for you. You're my last guy."

IT HAD BEEN CLOSE TO AN HOUR since the 60 Jayhawk had arrived at the sinking site. They'd just kept moving from one light to another light and another. Now they had almost a dozen men in oversized, waterlogged suits piled on top of one another inside the helicopter. It was almost 6:00 A.M. For most of the past hour, the cabin door had been kept open. The metal floor was a sheet of ice. The last fishermen to be brought up had been more hypothermic than the first. Some of them couldn't seem to hold themselves upright. Even though DeBolt had ordered two of the men to crowd onto the tiny console between the pilots, the back cabin was packed tight with bodies.

While he was operating the hoist, DeBolt focused his attention on what was going on outside the helicopter. But glancing over his shoulder now, the flight mech was alarmed by how cramped the cabin had become. One guy had his feet hanging partway out the open cabin door. There was water everywhere. Even one more body and there would be a risk of one of the survivors sliding out and ending up back in the Bering Sea.

"That's it," DeBolt told the pilots, as he pulled the basket into the helo with Evan Holmes inside. Never before in his career had DeBolt been in the position of even considering leaving anybody behind. Yet now he was sure that it wasn't safe to lift a single person more. The flight mech got the fishermen packed in as well as he could, then lowered the hoist line one more time for Starr-Hollow, who was floating alone in the waves below.

It was 6:00 a.m. when Brian McLaughlin heard the C-130 break in over the radio.

"Coast Guard rescue 6007, this is rescue 1705."

The Herc, McLaughlin realized with relief, must be overhead.

"1705, 6007. Good to hear from you," McLaughlin answered. "How you doing?"

They were good, the C-130 crew member answered. He asked for the helo's position.

"Roger," McLaughlin replied. "We're at 5, 3, 5, 3 north. 1, 6, 9, 5, 7 west. We've been keeping our guard with the *Alaskan* [sic] *Warrior,* they're Good Sam en route right now.

"We presently have seventeen people on board including our rescue swimmer," McLaughlin continued. "Or at this point, well, sixteen, we're about to pick our rescue swimmer up now. Then we're going to offload some of these passengers."

"6007, rescue 1705. Good copy, you have guard with *Alaska Warrior.* We'll relate to COMMSTA, and we will, we will take your guard at this time."

From McLaughlin's perspective, the back of the Jayhawk looked like a sea of red. The fishermen were so packed in, it was hard to tell where one body stopped and another began. Though the first few men they'd pulled up had been fairly energetic, the end of that big group included a couple guys who were severely hypothermic. Some of them needed to get warmed up right away.

McLaughlin knew from his earlier conversation with the *Munro* that District Command was recommending the helo bring their victims to Dutch Harbor. The search and rescue coordinators in Juneau had been in touch with personnel in the fishing port, who were ready to receive the survivors. McLaughlin's crew would be able to refuel there, then return to the scene. It would be the safest option for the men they already had, and for the crew.

McLaughlin remembered a couple times when a Coast Guard aircrew had been told there'd be fuel in Dutch. And when they got there? No gas. The Coast Guard wasn't supplying the fuel in the port; it was a civilian, a local guy. Sometimes he would say he'd be there. Then he wouldn't. More important, though, Dutch was such a long trip—150 miles one way. To McLaughlin and the rest of the Jayhawk team the decision was obvious—Dutch was out. It was just too far. All around them individual lights were still beating in the waves.

"Roger," McLaughlin answered to the Herc's offer to take the Jayhawk's guard.

"Can you guys find out where the *Munro*'s at?" McLaughlin asked. "We need to find out how far out they are. We have an option of dropping these survivors off to the *Warrior* but with these seas, it's just—wherever we put them is going to be hairy.

"If the *Munro*'s close, it would be ideal to bring them [there], but I think they're still going to be about sixty miles out."

"Roger that, 6007. Good copy. We'll . . . get that information for you."

McLaughlin had already been in radio contact with the *Alaska Warrior.* The FCA trawler was less than ten miles away. The pilot knew that the *Munro,* which had both fuel and trained medical personnel, was much, much farther. The Jayhawk was too big to land on either vessel.

With a dozen lights still flashing below, and every minute reducing those men's odds of survival, McLaughlin made a tough call: They'd attempt to lower their survivors to the *Warrior.*

THE CREW OF THE COAST GUARD's fixed-wing airplane, the Hercules C-130, was at twenty thousand feet when they began their descent. Pilots Matt Duben and Tommy Wallin calculated that they would save fuel—and time—by using the acceleration created by their descent to reach the disaster site. There were two needles on the Herc's airspeed indicator: a white one that displayed the plane's current speed and a red one that pointed to the aircraft's maximum allowable speed.

Duben and Wallin thought of themselves as good stewards of Coast Guard property. They weren't going to push the aircraft to its limits unless there was a damn good reason. Now they had one. People were in the water. Wallin kept his eye on those red and white needles, letting them hover right over each other. It was a technique called "flying the barber pole." Twenty thousand feet, ten thousand, five thousand. At three thousand feet they passed through the cloud layer—and into a snow squall.

Tiny flakes pounded against the plane's windshield. Then, more than two and a half hours after lifting off from Elmendorf, they saw it.

It was as if they were approaching a small city. More than a dozen tiny lights were pulsing eerily in the darkness. From fifteen hundred feet, the men couldn't tell which lights were people and which were rafts. Was there a boat? All they could see were the blinking strobes, signaling what they hoped were people still clinging to life in the waves below.

Chapter Nine

Sick at Sea

When the *Alaska Ranger* left Dutch Harbor early on Saturday afternoon, the *Alaska Warrior* was still tied up at the pier, its crew loading on the supplies needed for what might be a multiweek fishing trip. It was late afternoon by the time the *Warrior*'s deckhands untied the ship's lines and sailed out into Captains Bay.

Like the *Ranger,* the *Warrior* was a one-time Mississippi mud boat, an oil-rig supply ship that had been bought by the Fishing Company of Alaska and converted into a bottom trawler years earlier. The ship had about the same size crew as the *Ranger* and fished for the same species on the same fishing grounds. She was a slower boat, though, with a boxier design; the *Warrior*'s top speed was just 10 or 11 knots, as opposed to the *Ranger*'s 14 or 15.

Her captain was forty-six-year-old Scott Krey, a broad-jawed, blond-haired Seattle fisherman who'd been working in Alaska for twenty years, the last two as the captain of the *Warrior.* He was asleep when his first mate, Raymond Falante, called him in his stateroom, around 2:30 A.M. Ray had just heard from the mate on the *Alaska Spirit,* who had been called by David Silveira. The *Alaska Ranger* was sinking.

Scott knew most of the guys on the *Ranger;* he'd dealt with 80 percent of the crew at one time or another. Both Pete Jacobsen and David Silveira were on the ship as a favor to the company, Scott knew. Neither man had been happy to be put on there, but they were the types who were going to do their jobs.

The *Warrior* was already full throttle toward the *Ranger*'s reported position when Captain Scott got to the wheelhouse. They were about forty miles away—but still much closer than the *Spirit* or any of the other FCA ships.

MOST OF THE *WARRIOR* CREW HAD at least a couple of friends on the *Ranger.* It wasn't uncommon for Fishing Company of Alaska employees to move between boats. The *Warrior*'s first mate, Raymond Falante, had been the mate on the *Ranger* until late 2007. And the *Warrior*'s chief engineer, Ed Cook, had helped his younger brother Danny get a job with the FCA just a few months before. Now Dan was the chief engineer on the *Ranger.*

Ed's original idea had been that they'd share the position of chief on the *Warrior.* Ed was already sixty, Danny fifty-eight. They were ready to slow down, to spend a few more months a year somewhere other than Alaska. Ed, especially, wanted more time with his wife, Cindy. A couple years before, he'd been back at their modest riverside home in Washington's Cascade Mountains for just a week when the company called: They needed him to come back up immediately. The FCA would write Cindy a $1,000 "sorry" check to make up for it.

So Ed went back. It wasn't about the money, he'd just felt obligated. If he and Danny shared the same job, this type of thing wouldn't happen again, Ed thought. But it hadn't worked out the way he'd hoped. The company was desperate for qualified people. Right away, they put Danny on the *Ranger.*

Ed and Danny had grown up in San Diego, where their father was also a fishing boat engineer. They were the third and fourth of eight kids in a boisterous Irish Catholic family. When he was fourteen, Ed left school to join his dad on the tuna boats. A couple years later, Danny did the same. At seventeen, Ed enlisted in the Marines. Danny followed soon after. They both became sergeants and fought in Vietnam.

When they came home, Ed and Danny went back to tuna fishing out of San Diego. From the late 1960s, well into the 1980s, the Cook brothers traveled the world on tuna vessels. Grand and gleaming white, many of the tuna ships looked more like yachts than fishing boats. They sailed exclusively in warm waters: off the coast of Mexico and Chile and Peru, through the Panama Canal to the coast of Venezuela, across the Pacific to Hawaii, sometimes even as far as New Zealand or Puerto Rico.

Those were the good days. The few snapshots the brothers had taken in those years showed tan, muscular young men standing shirtless on the deck of a ship, their sun-bleached hair blowing in a tropical breeze. They'd dive and swim off the sides of the boat and pick coconuts and limes on the islands. Wherever the ship was tied up, they'd be able to find a good beach and maybe a great beach bar where most of the fights were over who would pick up the tab.

Along the way, Ed and Danny got their engineering licenses, just like their dad. Each moved steadily up the hierarchy of fishing life. The Cook men were part of a large community of fishermen from San Diego, America's top tuna port. For a long time, the fishing was good, and so was the money. But by the 1980s the industry was changing. Fishing had long been an international free-for-all. Now, most nations wanted to keep for themselves the profit and the jobs that their fishing grounds supplied. By the time the United States established its two-hundred-mile Exclusive Economic Zone (the EEZ) in 1976, many other nations were doing the same. The American tuna companies that for decades fished freely along the shores of foreign nations were now going out of business. Meanwhile, environmental regulations aimed at protecting marine mammals—particularly dolphins—had made it increasingly difficult to make a profit fishing for tuna in U.S. waters. San Diego fishermen began looking for work on foreign-flagged vessels. Some found it, but the jobs were scarce.

By the late 1980s, there was hardly any work left for what had long been a tight, proud community of San Diego tuna fishermen. But there were jobs in Alaska, increasingly in the remote Aleutian Island port of Dutch Harbor. Both brothers knew plenty of men from Southern California who were going north.

With foreign ships out of the newly established EEZ, the potential for American fishermen in Alaska was untapped. Adding to the boom were government incentives that encouraged the construction of new fishing boats or the conversion of preexisting ships into Bering Sea–ready vessels. And so far, the fishing grounds were delivering. There was crab, cod, pollack, and all sorts of groundfish—sole, mackerel, and perch. Big companies were running fleets of big ships and delivering the catch to plants right there in Dutch Harbor—or even processing it on board. Some of the boats in Alaska were hundreds of feet long. They all needed experienced engineers to keep them afloat.

The Cook brothers had worked on a number of different ships

over the years in Alaska. It had only been a few months that they'd both been employed by the FCA, though. Ed had mixed feelings about the company. The people were nice enough, but the *Warrior* was in poor repair compared to other ships he'd worked on. Right away, Ed had noticed that if the crew was inconvenienced by a watertight door, they would just tie it open when they were out at sea. That was goddamned lazy, he felt. And reckless. They were risking everyone's lives just to save the effort of opening and closing a door a few dozen times a day. It was the Japanese influence on the boats, Ed thought. It seemed to him like the Japanese didn't take safety seriously or respect American environmental regulations. He'd seen them dump dirty oil overboard at night.

Ed felt that the whole operation had low standards. There were constant problems with the sewage systems getting blocked up. On several occasions, Ed had seen raw sewage sloshing around on the hydraulic room floor. The damn pipes were in awful shape, covered with rust and held together with duct tape. Ed had tried to talk to the higher-ups in Seattle about the problems, and went as far as to send photos and videos he'd taken. He figured that, if they saw the boat's true condition, they'd take care of things. But there wasn't much of a reaction. As long as the fish were caught and processed and sold, no one seemed to care much about safety—or hygiene for that matter. He wouldn't want to eat those fish. For God's sake, Ed thought, the guys were pissing right on the floor of the factory. Ed had seen processors smoking on the assembly line. Sometimes the ash fell into the conveyor belt full of just-caught fish.

He'd tried to talk to management about it a couple times. "Don't worry about it, you're getting your paycheck aren't you?" he was told.

"How about a little pride in our work?" Ed countered, "In the product we're all out here breaking our backs for?"

The company's attitude didn't seem normal to Ed. He had worked for another big fishing company, Trident Seafoods, before signing up with FCA. One time a fish buyer had found a fly in a box of roe. Management was ready to launch a full *CSI*-style investigation. Was that an American fly or a Japanese fly? How did it get in that box? "This is high-quality product people are spending a lot of money for," Ed's boss had lectured the crew. It seemed to Ed that the FCA didn't care as much about quality. As long as they could sell it, it was good enough.

"ED. ED, THERE'S A PROBLEM on the *Ranger.* You should come up to the wheelhouse." It was the middle of the night when Ray Falante pounded on Ed's cabin door.

Ed pulled on his clothes as fast as he could. When he got up to the bridge, Captain Scott was at the controls.

"What's going on?" Ed asked.

"The *Ranger*'s sinking, Ed."

Scott told his chief that the *Ranger* may have dropped a rudder. From what he'd been told, the flooding probably started in the rudder room. It was spreading fast.

Ed looked out into the darkness. The deck was covered with ice. It was blowing hard with big seas.

Captain Scott knew how close the two brothers were. The two men even looked alike, for Christ's sake—they both had white beards, blue eyes, and round, soft faces that were always rosy red, even in the Alaskan winter. A lot of people had a hard time telling them apart, unless they were side by side. Then it was easy: Ed was the short one, just five foot ten to his brother Danny's six foot two.

Earlier in the day, just before the *Warrior* left Dutch Harbor, Captain Scott had been on the phone with the *Ranger.*

Ed was fueling the boat when Scott yelled down to him: "Hey, Ed."

"Yes, sir," the chief hollered back up to the bridge.

"Your brother called."

"He did? Well, what'd he say?"

Scott smiled. "To tell you that he loves you."

"Well," Ed had yelled back up. "If he calls again, sir, you tell him I love him, too!"

INSIDE THE NUMBER THREE LIFE RAFT the men were sitting in a circle with their backs against the inflated wall. The raft's waterbed-like floor was wet and the men were cold. Their hands and feet were numb. Outside, the waves pounded relentlessly against the raft. It felt like a roller-coaster they couldn't get off of. But at least they were alive. The Coast Guard knew where they were and had said they'd be back.

Fisheries observer Jay Vallee had been the first one in the raft. After the port-side number two raft—the one he was assigned to—was lost, there were more people on board than would fit in the remaining two rafts. Jay had told several crewmen that anyone who couldn't fit inside a raft should tie themselves off to the side. It was something he remembered from his safety training back in Anchorage. Better to be in the water and at least attached to a target large enough for rescuers to see than to be alone in the Bering Sea.

Jay had crossed from the port to the starboard side of the ship and approached the bow rail. The stern, starboard raft—the *Ranger*'s number three raft—had just been launched and was moving up the side of the ship, bolting toward the bow as the ship plowed backward into the sea. It looked empty. Jay was petrified, but somehow he managed to focus. It felt almost like an out-of-body experience. He took a breath and jumped twenty feet down the side of the boat—and right through the doorway of the tented shelter. He landed on his feet, then fell back on his butt with his legs pointed toward the middle of the raft. He dislocated his right ankle, but, amazingly, he'd done it.

Soon afterward, Jay had helped to pull in a few of the *Ranger*'s processors, including David Hull. Jay could see that David still had his laptop bag with him. You've got to be kidding me, Jay thought, as David tumbled into the raft. Soon there were ten of them inside. Every other person except for Jay had been submerged in the water first and was then pulled in. They were wetter than Jay—and colder. He looked around. People were quiet. They looked scared. Jay's suit fit, and he was dry. He had his PLB, which he knew was still signaling his position to satellites overhead. He was better off, Jay knew, than most of the guys he'd been working with for the past three weeks.

LIKE THE *RANGER*, THE *WARRIOR* SAILED with two federal fisheries observers every time it left port. Both Beth Dubofsky and Melissa Head were in their twenties and both had been working as observers since the previous summer. Beth and Melissa knew Gwen and Jay. All the observers assigned to FCA ships worked for the Anchorage-based observer contractor Saltwater Inc., whose employees stayed in the same bunkhouse when they were in Dutch Harbor. Now the *Warrior*'s female observers took charge of gathering supplies to treat potentially hypothermic survivors. The most important areas of the body for recirculating heat, they remembered, were the armpits and groin. The two women gathered blankets, and potatoes to warm in the galley's microwave. They'd hold them against the bodies of any rescued *Ranger* crewmen.

In the first couple hours after he was woken up, Captain Scott had talked with both Captain Pete and David Silveira multiple times by SAT phone. At first the men didn't sounded too panicked: "Come and get us, we're taking on water," the *Warrior*'s captain was told.

But as the night wore on, the calls from the foundering ship became more and more desperate. The *Warrior*'s captain was already running his ship at full speed toward the sinking site when he got a frantic call. It was about 4:20 A.M., and the *Ranger* had just lost power: "Hurry up! Hurry up! Get here as fast as you can. Now!"

Not long after, Scott heard the *Ranger*'s officers report that almost everyone was overboard—all but seven men. Some people hadn't made it into the rafts. Exactly how many was unclear.

Everyone in the *Warrior*'s wheelhouse could listen in on the transmission between the sinking ship and the Coast Guard. They knew that additional help was on the way, but it wasn't clear if it would come soon enough.

The *Warrior* was still an hour and a half away—fifteen or sixteen miles out—when Captain Scott spotted a blip on his radar screen.

Then, all of a sudden, the dot disappeared.

The *Ranger* was gone.

INSIDE THE NUMBER THREE LIFE RAFT, everyone was quiet. Along with nine other men, David Hull was sitting in a shallow puddle of freezing water that had accumulated on the raft's floor. The bottom part of his survival suit was flooded and he was cold. But his main concern had turned from his own life to the safety of those who stayed longest on the sinking ship—especially Captain Pete. Pete was the first officer David had worked for when he signed up with the FCA several years before. He liked and respected the captain, who would often come down to the factory to help the greenhorns pack fish. David knew that, like the other officers, Captain Pete had considered it his duty to stay on the vessel until the very last minute.

There wasn't much to say, and most of what the men were thinking didn't seem worth saying out loud. On the other side of the raft from David, cook Eric Haynes was examining the rips in his gloves. They tore when he was trying to pull the raft back toward the ship on its painter line. He had rope burns on his hands and a bloodied thumb. Eric had to keep his fingers moving so they wouldn't lock up from the cold. Every couple minutes, he pulled open his suit just below his chin and exhaled warm air inside. Outside, the seas seemed to be picking up. They know where we are, Eric thought. As long as we don't flip over, the Coast Guard will be back for us.

The twelve people inside the *Ranger*'s number one life raft weren't quite as lucky. Both rafts were the same make, designed for twenty people. Neither was overloaded. But the floor of the number one raft had filled with at least a foot of standing water. Though the people inside had tried, they couldn't get the raft's pump to work. One of the first men to reach the raft had ripped open the craft's survival pack, and now the flares and other emergency tools were all soaked, most of them lost under the mass of limbs obscured in the murky water. No one had a radio, no one seemed to have a working light. Worst of all, many of the dozen people inside had succumbed to seasickness.

Among them was Gwen Rains. She had activated her personal locator beacon at least two hours before. Jay Vallee had set his off at the same time, and soon after, there'd been a SAT phone call checking up on him. Gwen was in the wheelhouse the whole time, but there wasn't a second phone call. If someone called for Jay's beacon, Gwen wondered, why didn't they call for hers? Had they not picked up her signal? Or did they assume they were together?

Gwen looked at the men on the other side of the raft. She didn't know where Jay was now, but he certainly wasn't here. After only four days on the *Ranger,* she barely knew the names of anyone in the raft with her. There was Rodney, the assistant engineer, and the ship's steward. Most of the Japanese crew. The fish master wasn't among them, though. The last she'd seen of him he was sitting in the wheelhouse with a cigarette in his mouth, his survival suit falling open around his waist.

Gwen studied her beacon. Had she activated it correctly? Was the green light supposed to be blinking or the red one? She tried to pry the hard plastic cover off, but she couldn't get a good grip with her hands inside her suit's neoprene gloves. Finally, she ripped the plastic off with her mouth—and took a chip out of her front tooth in the process.

And yet she still couldn't tell if the PLB was working. One guy she didn't know came over from the far side of the raft and tried to help her. He took the beacon and looked it over. Everyone else was ignoring her. No one was talking, and many of the men had their eyes closed. The raft was pitching and jolting in the swells. It was already floating so low with all the seawater inside. What if the raft capsizes, Gwen worried. She had lost her strobe light while abandoning ship. It was on and blinking when she was on the *Ranger*'s deck. But once she was in the water, it was gone. The beacon might save her life. It was the best signaling device they had besides the raft itself, which would be a small target in stormy conditions in the middle of the Bering Sea.

Gwen could feel all kinds of stuff sloshing against her in the water. She knew that the survival pack in the raft would have had food rations and water, as well as a flashlight, an emergency blanket, and seasickness medication. There were plenty of people who could have used it. Gwen was violently seasick, and she wasn't the only one. There was no choice but to throw up straight into the life raft's standing water. It was humiliating, but she'd never felt so ill in her entire life. At one point, half the people in the raft were vomiting into the freezing, foot-deep water.

In the distance, Gwen could hear what sounded like the buzz of a helicopter. She guessed they'd focus on helping the men in the water first. Gwen just hoped they saw her raft as well.

INSIDE THE *WARRIOR*'S WHEELHOUSE, Scott Krey and Raymond Falante were working the radios. They'd been on with the Coast Guard cutter *Munro,* the Communications Station in Kodiak, and the pilot on board the helicopter that had taken off from St. Paul Island.

"Have you heard anything about my brother?" Ed Cook asked Captain Scott.

He hadn't, Scott told his chief. All he knew was that some guys were in life rafts. Others weren't.

Ed stared at the sea. It was so rough out, so cold. Danny's out there somewhere, he thought. He's out there in that pitch black night.

Ed was listening in as Captain Scott talked to one of the FCA's port engineers on the SAT phone: "We got twenty people in the water," Scott reported. "A few people are in the raft. The helicopter's on scene right now."

It's damn good luck that bird was up in St. Paul for crab season, Ed thought. Just really good luck.

Scott was still on the phone when Jayhawk Aircraft Commander Brian McLaughlin broke in over the VHF right around 6:00 A.M.

"*Warrior,* Coast Guard copter 6007. Can you give me your present position, sir?"

"Hang on, I gotta go," Scott told the company engineer as he put down the phone and picked up the VHF.

"My position: 5, 3 degrees, 5, 3.7 north. 1, 6, 9 degrees, 4, 8.8 minutes west," the captain relayed.

"Roger, Captain. Copy 5, 3, 5, 3.7. 1, 6, 9, 4, 8.8. Is that right?"

"That's roger."

There was a long pause, and then the 60 aircraft commander hailed the *Warrior* once again.

"*Alaska Warrior,* Rescue 6007."

"*Alaska Warrior,*" Scott responded.

"All right Captain, we're headed toward you," McLaughlin said. "How's your ship riding right now?"

"It's riding pretty good, pretty good."

"Okay, we're gonna drop some people off. We've got, I believe, thirteen on board right now. We're gonna have to get them off as quickly as possible," McLaughlin told the captain.

In the twenty years that he'd been fishing in Alaska, Captain Scott had seen plenty of injured crewmen airlifted off fishing boats by the Coast Guard. This would be trickier, but if the pilot thought it would work, it was worth a try.

"I'm going to need your deckhands to help us out a lot," McLaughlin told the *Warrior*'s captain. "A lot of these people are nonresponsive, and/or holding on to the basket when we get them in there. You may have to be a little rough with them, but we need you to get them out of the basket as quickly as possible, so we can get them out and get back to the people out here. Out, copy?"

"Roger," Scott answered. "Copy."

If the chopper was dropping guys on their deck, the crew

would need instructions, Ed knew. They would need to know that the basket had to touch the deck before they grabbed onto it. Otherwise, they could get a terrible shock from all the static buildup from the copter's rotors.

"Captain, you want me to carry a message back for ya?" Ed asked.

"Tell them some of them are nonresponsive," Scott said. "The basket drops, drag 'em up. *Drag 'em up*. Don't worry about hurting them, get them out of there."

From inside the *Warrior*'s wheelhouse, the men could hear the rotors approaching.

"There he is," Ed said, just before McLaughlin's voice broke through again on the VHF radio.

"*Alaska Warrior,* Coast Guard 6007. Do you think you'll be able to run downswell? Would your ride be any better at that point?"

"Yes it would," Scott said. "I can turn."

"All right, if you could run it downswell that'd probably be better for us as well to get on top of you."

The wind and waves were coming from the northeast. Now Captain Scott steered the *Warrior* toward the southwest.

"How about our speed?" the captain asked.

"Pretty much keep it at clutch speed. As long as you have steerage, as long as you have control, you can't be too slow for us. . . ."

It was 6:10 a.m. as the Jayhawk approached the *Warrior*'s port side. The clap of the rotors grew to a thunderous roar as the machine settled into a hover over the trawler's stern. Outside, it was still spitting snow and blowing hard.

When the door to the helicopter slid open, the crew on the *Warrior*'s trawl deck could make out a mass of red inside. They watched as the metal basket dropped out of the cabin, and slowly descended toward them with Evan Holmes inside.

He was terrified.

The *Ranger*'s factory manager had been the last person the Jayhawk had lifted out of the water. The cabin was already so crowded that flight mechanic Rob DeBolt told Evan to stay right inside the basket. There was nowhere else for him to sit.

Evan had been shocked at how packed the chopper was; he didn't know a helicopter could hold that many people. With all the noise, he couldn't hear what was going on. But he knew he was the last one. He'd only been in the Jayhawk for a couple minutes when DeBolt leaned over and yelled in his ear: "Hey, we're going to drop you to the *Warrior,*" the Coastie told him. "Stay in the basket. Hold on."

"What?" Evan said. "No! Put somebody else first." He turned, and yelled to one of his crewmates. "You go! Tell me how it works out," he tried to joke.

But it was obvious to Evan that it wasn't up to him. He was going first.

As he was lifted out of the cabin, Evan could see the *Warrior*'s huge trawl net strewn out over the deck. There were buoys everywhere. The gantry seemed way too close as the two-hundred-foot ship pitched and rolled in the waves. Even the crew on the ship's deck looked like they were barely holding on. Evan could see a few of their faces. He knew some of the guys. There was a big Samoan dude he liked. *Oh, man, I hope he catches me. I do not want to smack the boat,* Evan thought.

Evan cowered inside as the basket swung like a pendulum above the icy deck. He hugged his arms around his legs, trying to keep fully within the metal box. *Jesus Christ, I just had my* life saved, and now I'm gonna die getting banged against the goddamned *Warrior,* he thought.

The deck grew steadily closer. On the flight to the *Warrior,* some of the *Ranger*'s crew members felt good enough to joke around a little in the back of the helo. "They'll probably make us work," someone had said. Evan didn't doubt it. He didn't want to get on the *Warrior.* He was halfway down when all of a sudden the basket started rising again. They were bringing him back up, Evan realized with relief. They'd changed their minds.

Evan reached the cabin door, and DeBolt steadied the basket against the edge.

Then, horribly, he was going back down. They were trying again. Evan closed his eyes. He didn't want to see it coming; there was so much rigging, so much gear. He could hit the ship, the wheelhouse, one of the boat's sharp, pointy antennas. The basket was spinning; he was spinning. He was scared shitless. There was water coming from all directions. It was so windy, Evan couldn't tell if it was raining or if water was just being blown up from the ocean or down from the helicopter. The rotor wash was so powerful he couldn't look up.

I'm more likely to get killed right now than I was back in the water, Evan thought. This seemed like a bad idea.

Then, suddenly he was moving back toward the chopper. DeBolt steadied the basket just below the open door while the pilots repositioned the helo over the front of the ship. Evan saw the *Warrior*'s crew running toward the bow.

Evan looked up toward the pilots. Through the aircraft window, he could see that one of them was shaking his head and slicing his hand across his throat. Moments later, the basket was pulled back into the cabin.

"Don't do that again!" Evan shouted to the flight mech as soon as he was inside.

"Don't worry, man," DeBolt yelled back. "It's not going to work."

The 60 Jayhawk had been hovering over the *Warrior* for less than five minutes when, with the consensus of his crew, McLaughlin made the second tough call of the night: Lowering the fishermen to the *Warrior* was just too dangerous.

They would have to offload the men to the cutter *Munro* instead.

Chapter Ten

Man Down

As the giant *Munro* lurched through the waves, the tiny 65 Dolphin helicopter clung to the flight deck on the ship's stern. The helo resembled an unwieldy piece of furniture cinched to the roof of a car on a potholed road. It was perfectly secure—but still looked precarious.

Pilots TJ Schmitz and Greg Gedemer zipped up their orange dry suits and pulled on their visored helmets. Bracing against 35-knot winds, they ran from the hangar out to the aircraft. The flight deck netting whipped violently in the wind as the men walked around the helicopter, making sure the aircraft hadn't built up too much excess ice. Then the pilots climbed into the cockpit and started her up as rescue swimmer Abe Heller and flight mechanic Al Musgrave buckled up in back.

"The limit light is on," Gedemer told Schmitz. "It's flashing on and off."

An indicator on the helo's instrument panel warned that the conditions outside the aircraft were out of limits for takeoff.

"Yeah, that's because the blades are flopping all over the place," Schmitz said.

The wind across the *Munro*'s flight deck was so strong that the helicopter's computers had determined the aircraft was already at limits. They'd have to do a high-wind start.

The *Munro* had turned to secure the best launch course. In the engine room, the ship's engineers had stopped the vessel's high-speed turbines and were back on the diesel engines. They slowed the ship to about 10 knots and pointed the bow straight into the swells. Once the pilots were in the helicopter and hooked up to the ICS, they could communicate with Erin Lopez and the crew in Combat, who were in direct communication with the engine room. Up on the bridge, Captain Craig Lloyd was also looped in.

When their landing signal officer (LSO) gave the okay, four tie-down crew ran out onto the flight deck, hunched down against the powerful blow of the 65's rotors, and uncinched the wide canvas straps that held the helicopter tight to the deck. Bundled up in bulky, insulated blue jumpsuits, matching helmets, and vest-style PFDs, the tie-down crew was easy to distinguish from the aircrew. The *Munro*'s officers had taken to calling them the "blueberries." The captain and crew could watch the action on a series of black-and-white video screens on the bridge. The sequence looked like a well-choreographed dance, the 65 helicopter the prima ballerina among a troupe of little scurrying mice.

Everyone knew that the conditions were right on the edge of limits. Or, more accurately, every few minutes, there was a minute or so that was in limits. If this had been a training exercise, it'd have been canceled. But in life-or-death situations, the call is up to the commanding officer, and Captain Lloyd trusted his pilots. From the moment Schmitz arrived in Combat and heard the details of the case, he'd felt confident they would be able to launch. Now Schmitz was in the right seat with Gedemer at his left. He studied the incoming swells through his night vision goggles. They were close to twenty-footers, but rolling in at a pretty steady pace.

Schmitz had already briefed the crew on the takeoff conditions: "This is the deal," he told them. "We're going to overtorque the airframe when we take off. As long as we don't pull more than eleven point eight, we can continue on in the mission."

Coast Guard regulations lay out stricter operating standards at night than for daytime flights. In the dark, anything over 4-degree pitch, 5-degree roll is considered out of limits. But when a mission involves an opportunity to save a life, the men are authorized to go beyond those limits—even at the risk of damaging the aircraft. "Warranted effort," the regulations call it.

Back in Combat, Schmitz had briefed Captain Lloyd on his plan. He would load on 1,750 pounds of fuel, several hundred pounds more than normal. Though the 65 has a powerful engine, its gearbox is relatively weak, which means that the weight of a full load of fuel makes it difficult, if not impossible, for the helo to maintain a stable hover—a more power-consuming maneuver than forward flight. In any case, they'd most likely burn off a good quarter of their fuel load just to reach the scene of the sinking. Schmitz told Captain Lloyd that he was expecting to overtorque, or stress, the main rotorhead on takeoff, but wouldn't push it so far that he'd sacrifice the ability to keep flying. "In other words, I'm gonna break it, but I'm not gonna break it so bad I have to land right away," the pilot had told the captain.

Now Schmitz watched the waves. A set of big swells was rolling in perpendicular to the cutter's bow. Under normal conditions, a pilot would launch at a lull between waves, waiting for the calmest moment to lift from the deck. Not tonight.

"6566, ready to take off," the crew heard up on the bridge.

Schmitz waited until the next wave pitched the bow into the air, and just as it began the sharp fall that would buck up the stern, Schmitz pulled the 9,000-pound bird off the deck, using the momentum of the lurching ship to catapult the helo into the air.

Gedemer called out the torque as Schmitz slid the aircraft to the ship's left side. The helo's official torque limit at takeoff is 10.3; Gedemer's highest number was 9.9. They were golden—they hadn't overtorqued after all.

Schmitz's unconventional maneuver had worked.

It was just a couple minutes before 6:00 A.M. as the crew marked the *Munro*'s position and set off toward the last known coordinates of the *Alaska Ranger.*

JIM MADRUGA HAD TO PEE—BAD. He'd been in the water for what seemed to him like three or four hours. What else could he do? When he finally let go, the warm stream felt so good. He was cold, but, except for the urine inside his Gumby suit, dry.

His suit fit, the seals had held, and he felt alert. He was clearly doing better than the fisherman who had been floating with him for the past couple hours. The guy was out of his mind with fear. Jim had tried to calm him down, but hadn't been able to help much. Every time they heard an aircraft overhead, the guy started screaming.

"Save your strength," Jim told the younger man. "They'll be coming for us, but they have to get the other people first."

Jim was the *Ranger*'s second assistant engineer. He was fifty-nine years old, and like many other men of his generation, he was a former San Diego tuna fishermen who had made a second fishing career for himself up north. He'd headed straight to the wheelhouse after being woken up that morning. His chief, Dan Cook, was up there and had already concluded that the ship wasn't salvageable.

"Abandon ship" was all Dan had said to him.

Jim and Dan went way back. They'd both started fishing in Alaska more than a decade before. For a few years, both men worked for Trident, one of the biggest fishing companies in Alaska. One year, they'd sailed one of Trident's old ships to India to be scrapped. It was a skeleton crew on an adventure, and Jim and Dan got to know each other well. When they reached India, they drove the boat right up onto the beach at full steam. The carcasses of other ships littered the sandy expanse, and they watched as the Indian scrap crew went at it. They looked so poor, those skinny little guys wear only sandals and skirts to crawl around on a heap of rusty metal like that.

A few years later, both men found themselves working for the same company again. Dan's brother Ed had been with FCA for a few years, as had David Silveira, a cousin of Jim's from San Diego. That's the way it worked in fishing—the same guys again and again over the years.

A handful of men had been in the wheelhouse when the last call was made to the Coast Guard, and then everyone got out. By that time, the water was almost to the wheelhouse door. Jim hadn't seen where everyone else entered the water. He was alone with Dan. The chief engineer had been in poor health. They were both nudging sixty-years-old, but to Jim, Dan seemed like he could be a decade older.

Dan wanted to wait until the very last moment to get off

the ship. Jim figured his friend was thinking that if they waited longer, they wouldn't be in the water as long before help arrived. He wanted to stay to make sure Dan got off safely.

The two men waited until there was literally no choice. Then they jumped off the boat together.

"Dan, try to stay with me!" Jim yelled. The wind seemed to be blowing 50 knots and the waves were at least twenty feet, and breaking. Jim watched helplessly as Dan was pushed away by the waves.

After a few minutes, Jim drifted up next to some fishing net and buoys—debris from the ship's deck. He grabbed on. He'd been floating there for about half an hour before he saw Byron, one of the new kids, a Hispanic guy who'd been on the boat just a few days. Jim hadn't talked to him too much, but he'd sat next to him at lunch a few days before. He remembered Byron's name. Jim also remembered that this was his first time on a fishing boat.

The engineer grabbed the younger man and pulled him into the net. They hadn't been in the water that long, but already the kid seemed to be going into shock. "What's the matter, Byron?" Jim asked.

"I'm so cold," Byron cried. "I'm so cold."

Jim pulled Byron closer and put his arm around him. He tried to keep him talking to get his mind off things. The younger man spoke a little about his family, a wife and two young daughters back in California. But mostly he just kept mumbling about how cold he was, and about all the water that had filled his Gumby suit.

Jim just held on to him, and they stayed with the net. It gave them a little buoyancy, and Jim figured it might be easier to see from the air.

"Help me! Help me!" In the distance, Jim thought he heard Dan Cook yelling. At least an hour had passed since they'd abandoned ship and Jim could barely make out his friend's screams over the wind. He scanned the waves. For a moment, he thought he saw Dan floating on his back about fifty feet away. But there was no way he could get to him. Before long, Dan had drifted out of sight again.

Eventually, Jim saw what looked like a Coast Guard plane overhead. Byron must have seen it too because he began to yell.

"Just save your breath, man," Jim told him. "They can't hear you."

Jim's strobe light was out. It had worked on the ship, but after he'd been in the water for a while, he noticed it had gone dark. Byron's was working fine, though, so Jim knew they were visible.

Was there a boat in the distance? Jim could see a bright light right on the horizon. At first it seemed to be getting closer, but then it stopped. *Maybe they're getting people out of the water,* Jim thought.

A while later he saw a helicopter in the distance.

Again Byron started yelling, but Jim pleaded with him.

"They know we're here," he said. "They will come eventually."

AIRCRAFT COMMANDER TJ SCHMITZ pulled the 65 Dolphin up to about five hundred feet and started south toward the last known coordinates of the *Alaska Ranger.* Even with the tailwind, the seventy-mile journey would take them about forty-five minutes. They'd heard some chatter over the radio from the Coast Guard rescuers already on scene. It sounded like the larger aircraft was at capacity, and had left quite a few people behind in the water.

Schmitz started talking strategy. He was expecting the worst. By the time they got there, some of these people would have already been in the water for close to three hours. He knew from experience that, even in survival suits, many people couldn't make it in cold water for more than two.

Do we pick up people who might be dead? Schmitz thought. Or do we pick up people who seem most responsive? He posed the question to the rest of the crew. In Schmitz's opinion, the best course would be to focus on the most responsive people first.

"Even though they're not responsive, they may not be dead," rescue swimmer Abe Heller noted from the back of the helo.

"Yeah, but we only have so much room," Schmitz said. "Depending on how many people are in the water, you know, who do you go for first?"

When the 65 Dolphin was about fifteen miles north of the site, Gedemer spotted the larger 60 Jayhawk helicopter to their west.

He picked up the radio: "6007, this is rescue 6566."

Brian McLaughlin told the Dolphin's crew that there were two rafts holding survivors—and that at least a dozen or more people were still in the water.

"The survivors are getting less and less responsive," McLaughlin reported.

Based on the inflection in McLaughlin's voice, Schmitz anticipated a grim scene. He knew their tiny helicopter couldn't possibly get even half of the people out there in one load.

"Okay, we're going to need to do this as fast as we can," Schmitz told the rest of the men.

Schmitz had spent his previous four-year tour in the Great Lakes, where he had plenty of experience with hypothermic victims. The rescue crews had often used a procedure called the "hypothermic double lift." In the later stages of hypothermia, a person's blood collects near the heart and vital organs. If the victim is suddenly lifted upright from the water, there's a risk that this blood will rush from the torso into the legs, causing heart failure. Because of that risk, professional rescuers are taught to keep hypothermic victims in a horizontal position. Coast Guard helicopter rescuers are trained to use a double-harness system. The regular quick strop is fastened around the victim's knees, while a second strop—a larger, older model, sometimes called the "horse collar"—is secured under their armpits, then tightened around their chest. Ideally, the hypothermic victim will be raised with knees and chest at about the same level, like a bride being carried over the threshold.

The crews had been trained that the two-strop method was the best for someone with hypothermia, but they knew that it was an inconvenient lift. It was time-consuming to get a victim settled securely into the two-strap setup, even in calm conditions when the survivor was alert and cooperative. With severely hypothermic survivors? In high seas? In the dark? It might eat up fifteen to twenty minutes per person.

We don't have that much time, Schmitz thought.

In the back of the helo, Musgrave and Heller were thinking the same thing: the basket. With the metal rescue basket, they'd be able to raise the hypothermic fishermen in a seated position and minimize the risk of heart failure. The basket would also be faster because there were no straps to secure. With his dry gloves on, Heller knew he wouldn't have much mobility in his fingers. It'd be tough to fiddle with the buckles on the straps and the clip on the hoist. With the basket, none of that would be a problem.

Heller had been on two long patrols since his arrival in Kodiak's ALPAT shop two years before, but he'd never launched in conditions like this. Back on the ship, Aircraft Commander Schmitz had pulled the swimmer aside. This would be a "load and go" mission. It was clear people were in the water. Hopefully most were in life rafts, and it was likely they'd be lifting people from rafts up to the helo. There'd almost certainly be more people than they could take in one trip.

"If it's all right, I may want to leave you in a raft out there," the veteran pilot had said to the twenty-three-year-old swimmer.

Heller was ready to do whatever was necessary. He went back to his rack and bundled up in everything he had. He knew that the more layers he wore under his dry suit, the less dexterity he'd have in the water. But it was cold, and there was a good chance he could be in the Bering for hours. The tradeoff was clear. His first layer was long underwear made of Nomex fleece, similar to Polar fleece, but with fire retardant built in. Over the long johns, the swimmer wore another pair of fleece pants, two more shirts, a fleece unitard—a "uni" the Coasties called it—and then a heavy coat and wool socks. Last, he layered on his orange dry suit and his neoprene and rubber boots. Depending on the situation, he might have chosen wet gloves. Those form-fitting gloves offer more dexterity—but less warmth. But this morning he couldn't risk frozen fingers. He grabbed the dry gloves. Limited mobility would be better than none.

Like rescue swimmer O'Brien Starr-Hollow, Heller had landed in Kodiak on his first assignment as a rescue swimmer. Historically it was rare to end up in Kodiak for your first billet as a swimmer, and even rarer to end up in Alaska Patrol, the ALPAT shop. Heller figured maybe the guy doing the assigning that month didn't know the guidelines. In any case, it'd worked out well for him. He liked Alaska. He'd grown up in Wyoming in a Navy family, graduated high school in 2003, and joined the Coast Guard the following December. All his life, Heller had been interested in aviation. His dad worked for the Wyoming Bureau of Land Management as a range conservationist. He dealt a lot with helicopters, helping to manage the aviation side of wildland firefighting. Heller researched the Coast Guard's three aviation rates, and Aviation Survival Technician—AST, or rescue swimmer—sounded like the most fun.

He went through boot camp, made it to A School a couple years in, and failed out two weeks later. It was the "rear release" test that got him. It's a drill meant to prepare the swimmer for managing a panicking survivor. An instructor grabs hold of the swimmer from behind, and the swimmer has to take him underwater, wrestle him off, and come up with the instructor in tow.

Heller didn't know why he failed. He'd done the rear release perfectly in training several times. But each time it counted, he couldn't pull it off. He was one of the five of ten in his original class who didn't make it through the course. If Heller had raised his hand and said "I quit," there would've been no second chance. But if a student is injured or just has trouble with a specific skill, he or she may be permitted to try again.

Heller spent the rest of the fall of 2005 in the airman program in Elizabeth City. He prepared for two more months—more physical training ("PT" as the military calls exercise programs), more pool time. His next class started with nine students. Four months later, seven graduated, Heller among them. After A School, Heller spent three weeks at EMT school in Petaluma, California, and then went up to Kodiak. He'd been there since the spring of 2006 and had a little apartment in town. He'd made lots of friends and gotten in tons of snowboarding on Pyramid Mountain. He'd been involved in a couple of SAR cases.

His first was a medevac—a slip-and-fall victim off an 850-foot container ship. They flew out, Heller was hoisted down to the ship and put the injured crewman on a litter. They hoisted him up and flew him to Dutch Harbor. The whole mission took about an hour. A piece of cake. The next summer, Heller was deployed to Cordova and ended up on another medevac. It was a four-wheeler accident on a remote beach. A kid was getting towed on a sled behind an ATV (all-terrain vehicle) and was tossed off. The Coasties airlifted the kid and brought him back to Cordova for medical treatment.

Heller had only been in the water on a real case one time. Schmitz had been one of the pilots on that case, too. It was a small fishing boat, just around thirty feet. The vessel got too close to shore, got caught in the breakers, and ended up capsizing. The helicopter crew found it at around midnight, lying on its side inside the surf zone, getting knocked around by the waves. The pilots landed the helo on the beach and Heller waded out to the boat. He swam around, looking, but there was no one there to save. The fisherman's body eventually washed up on a nearby island. Investigators guessed he'd fallen off the boat before it even got caught up in the surf. The Coast Guard had been too late; there hadn't even been a chance to help him.

SCHMITZ AND HIS CREW WERE about three miles out from the scene when they spotted the strobes. There were about ten of them, flashing on and off across almost a square mile of ocean. As they got closer, Gedemer saw that the two brightest lights came from the life rafts.

"Let's concentrate on the people as far away from the others as we can," Schmitz said.

He could see three lights to the northwest that were off on their own. *Those are probably the people who've been in the water the longest*, Schmitz thought. *That's where we'll begin.* The pilots brought the 65 into a hover over a single strobe while flight mechanic Al Musgrave attached the basket to the hoist's talon hook, and motioned for Heller to climb in.

"Basket at the door," Musgrave said as he contracted the cable to pull the basket from the floor of the aircraft up and out of the helo. "Basket going down." Within seconds, the compartment hit the swells below. Heller climbed out and swam toward the fisherman alone in the ocean.

"How're you doing?" Heller yelled as he grabbed on to the man.

"Cold as fuck!" the fisherman screamed back.

All right, Heller thought. This one was probably going to be okay.

The swimmer had a small green chemical light attached to his mask, similar to the glow sticks sold to kids at country fairs. In dark conditions, it was common to use the light in place of simple hand signals to indicate ready for pickup. Heller pulled the light off his mask and waved it above his head to signal to Musgrave to drop the basket. It took just a couple of minutes for Heller to load the fisherman in and to give Musgrave the thumbs-up to raise the compartment.

As the flight mech began the hoist, Heller hung on to the bottom of the metal basket, helping to steady it and center it under the aircraft as the lift began. Once he'd been pulled a few feet clear of the water, Heller dropped back to the sea and watched the basket rise another fifty feet to the open door. Musgrave pulled it inside and dumped it sideways so the fisherman could crawl out. Then he sent the basket back down for Heller.

The rescue swimmer climbed in, the flight mech raised him up, and they moved on to the next light with the basket steadied right outside the open helo door. It was just one hundred or so yards away. That pickup went smoothly, as did the next rescue, but already the helo's cabin seemed packed.

"I think I have room for one more," Musgrave reported to the pilots.

"You're gonna have to stack 'em up," Schmitz said. "We're going to stay on scene until we run out of gas, and we have to go back."

The first man who'd been pulled into the helo had inexplicably started stripping off his survival suit. The second survivor followed his lead but got stuck. He was kicking, motioning for Musgrave to help him pull the suit off the lower half of his body. The flight mech thought it would be better if the fishermen just left their suits on, but there was no time to argue. He grabbed a webbing cutter from inside the doorframe and pulled it right through the zipper of the man's suit. Musgrave sliced from the fisherman's chest all the way down one leg, like he was opening up a fish. Then he threw the guy a wool blanket and directed both him and his buddy to the back of the cabin, where the rescue basket is normally stored.

The nose of the helo suddenly jerked upward.

"Holy shit! What's going on back there?" Schmitz asked through the ICS.

"The guys are climbing in the back," Musgrave answered.

"All right, put the next guy behind Greg, against the door," Schmitz replied.

The men's movement had affected the helo's center of gravity, but at least they'd made room for more survivors. Heller was still in the basket against the open aircraft door as Schmitz circled around to the south toward a clump of lights in the water. The rescuers saw four men locked arm-in-arm, like a human chain. The two on the ends were waving their free hands. They looked like they were in pretty good shape.

Then Schmitz saw another strobe about a hundred yards away.

"There's another survivor off the nose. Let's go get him first," the pilot said.

They only had room for a couple more people in the helicopter. Better to get those who were off on their own, Schmitz thought.

When they got closer, though, the pilot could see that it was two men tangled up in a bunch of netting and buoys with one strobe light between them. Jim Madruga was waving his arms; Byron Carrillo was just floating.

They came into a hover beside the net, and Musgrave placed Heller a good hundred feet from the debris pile. As the flight mech drew the empty basket back to the aircraft, Heller swam up to the net. Byron was lying limp in the water with the webbing all around him, his strobe light flickering in the darkness

"Take him first," Jim yelled. "He's in really bad shape."

Byron was reaching for Heller, but the rescuer swam around behind him to get a safe grip. Heller wanted to avoid any struggle with the fisherman; controlling him from behind was the best way to avoid any problem. He asked Byron how he was doing.

With the rotor noise and the fisherman's delirium, Heller couldn't make out what he was saying. But the fact that he was speaking meant something.

Byron had his hand jammed under the net, and Heller struggled to pry open his fingers.

"You gotta let go!" Jim yelled. "You've got to let go!"

Finally, Heller broke Byron free, and began to drag him back out of the debris to a safe hoisting location.

In A School, Heller had been drilled in a method to clear rope, netting, or any other refuse from a submerged survivor. The swimmers were taught to go underwater, put their hands on the survivor's spine, and then walk their hands all the way down the victim's body, looking and feeling for debris. The technique was called a "spinal highway."

Byron was clear, but as Heller dragged him away from the

net, he could feel a piece of line snagged on his own fin. Heller used one hand to steady Byron as he reached down to clear off the debris. As the swimmer was working to free himself, a wave knocked the fisherman facedown in the water. Heller looked up to see that Byron wasn't righting himself.

Crap, Heller thought. This guy is so far gone he's not even capable of keeping his own face out of the water.

He grabbed onto Byron again, pulled him upright and swam with him away from the wreckage.

When Heller had the fisherman away from the debris field, he signaled for the basket, which was brought down in seconds.

At the sight of it, Byron seemed to snap to attention.

He knows what the basket is, Heller thought. The basket is life. He tried to maneuver Byron inside. He pushed him into the basket, but then Byron would change his position and end up crossways, with feet coming out one end, head coming out the other. Heller yanked him out and tried again. When he got Byron out of the basket, though, the fisherman wouldn't let go of the metal bars. Heller was struggling to get the compartment upright again while Byron was pulling it sideways into the waves, flailing in a panic. Heller fought with the man for at least ten minutes.

Heller knew that a hypothermic person often becomes irrational and can have symptoms similar to being drunk—loss of bodily control, slurred words, inability to focus, or pay attention to instructions. It wasn't the man's fault, but it was still frustrating. Every second it took to wrestle this one person was a second the rescuers could have used to pull someone else out of the water.

Up in the cabin, Musgrave was watching Heller struggle with the fisherman. Since he'd dropped the swimmer into the water fifteen minutes earlier, the flight mech had lost sight of Heller three times. It was called "losing target," and it was something the aircrew never wanted to happen. Disorienting snow squalls were blowing through the area and the waves were big and irregular. The swells kept pushing the swimmer and fisherman underneath the helicopter, which meant that Musgrave would have to instruct Schmitz to reposition the helo just so he could see what was going on.

Musgrave could hear the pilots talking about fuel. They were getting low, burning through their gas faster than usual. Hovering used up fuel a lot faster than forward flight did, especially with a full cabin like the one they had now.

Musgrave didn't want to let out too much slack on the cable. He couldn't risk getting it wrapped around someone's leg or neck. He let out what he thought was necessary, but still the basket was jerked out of the water a few times. They'd been at it for so long.

Finally, Musgrave looked down to see Heller and Byron centered between swells. The basket was plumb beneath the helo. It looked like if he pulled the basket out of the water, the fisherman would drop down to the basket floor where he needed to be. Musgrave began the hoist.

Down below, Heller was still working to get Byron seated properly when, all of a sudden, the basket lifted above the surface and started rising. Heller hadn't given the signal to lift the basket, but now it looked like Byron was in there pretty good. The swimmer felt relieved as he watched the basket rise toward the helo and saw Byron seem to slump down into the compartment.

The basket was fifteen feet above the waves when Heller turned and began swimming back toward the debris field, where Jim Madruga was waiting.

The older fisherman was still floating alongside the net.

Finally, he thought, as Byron was pulled up out of the waves. Jim watched the basket rise about twenty feet above the surface, then spotted the rescue swimmer coming back toward him. It felt like it had been a long time since the Coast Guard swimmer pried Byron from the net. Now it was Jim's turn.

From the open door of the helicopter, Musgrave saw Byron drop down a little bit inside the metal basket. Everything looked good. The flight mech watched his swimmer turn and start moving toward the next survivor.

He continued with the hoist.

"The survivor is in the basket," Musgrave reported into the ICS. "The basket is out of the water. The basket is above the water."

Halfway up, Musgrave saw that Byron seemed to have lifted himself onto the rail. He looked like he was actually sitting on a shorter edge of the rectangular basket; his butt was on one corner and his feet were hanging over the adjacent side. Byron had his arms wrapped around the bales. He wasn't where he should be, but he still looked relatively stable.

But with the basket moving up, more than halfway there, Byron slipped. From above, it looked like the lower part of his body was now outside of the compartment.

"The survivor is hanging from the basket," Musgrave announced.

"What?" Schmitz said.

"What did he say?" Gedemer asked.

Schmitz was confused, but from the right seat, he couldn't see what was going on in back. The hoist was Musgrave's show and he was the only one to see the basket reach the cabin door—with Byron hanging by his armpits from its side.

He seemed huge, as if he were seven and a half feet tall. Most of his body was outside the basket, with his legs hanging straight down below the bottom of the basket floor—and between the bottom of the basket and the helicopter. Byron's position made it impossible for Musgrave to pull the basket into the cabin.

Instead, he brought the hoist in as far as he could without getting Byron wedged up against the aircraft. He started trying to haul Byron into the helicopter.

Musgrave had never been in a situation where he couldn't manhandle somebody into the cabin, but he could barely budge Byron's legs. Kneeling at the doorway, he was just about at eye level with him. Byron's red neoprene hood was up, but quite a bit of black hair was hanging out. With the suit's mouth flap fastened, all Musgrave could see of the fisherman was his eyes and the bridge of his nose. There was no point in trying to say anything over the roar of the rotors.

Musgrave looked into Byron's eyes and saw that his face was frozen in terror.

All Byron had to do was move his legs a little bit, but he wasn't helping at all. His suit is full of water, Musgrave realized. He probably weighs 500 pounds. Musgrave reached back for a knife that was attached to the side of the cabin. He'd slice open the neoprene legs and get the water out of the suit. Then he'd be able to pull the guy in.

The knife was just a couple feet away, but in the moment that Musgrave moved to grab it, Byron slipped again. When Musgrave turned back to the open door, the fisherman was hanging by his elbows from the edge of the basket. Musgrave grabbed him, and pulled as hard as he could. But it was only two or three seconds before Byron let go.

He slipped out of Musgrave's arms, plunging into the sea forty feet below.

Moments later, Schmitz heard the mechanic's voice.

"We lost him. We lost him," Musgrave repeated.

"We lost who?" At first, Schmitz thought that Musgrave said, "We lost them."

Though Schmitz and Gedemer couldn't see what was going on back in the cabin, they'd known from Musgrave's silence when the basket reached the helicopter that something wasn't going right.

"The survivor," Musgrave said. "He's gone."

Schmitz could see the man's light blinking in the water below. For an instant, he thought he saw him move his arms in the waves.

"He's okay! He's moving," the pilot said.

But seconds later a heart-wrenching reality set in: "Never mind. He's facedown."

CHAPTER ELEVEN

Out of the Cold

After the Jayhawk's failed attempt to lower their survivor to the *Alaska Warrior*, there was no doubt in Brian McLaughlin's mind that the next best course of action was to bring their cabin-load of fishermen to the *Munro*. The aircraft commander kept thinking about those blinking lights. They needed to get back to the people they'd left behind. The *Munro* and its crew were reliable, and they were right there. It was a more complicated, more dangerous approach than delivering the survivors to Dutch Harbor, but there were lives at stake. Many lives. As they got closer to the cutter, McLaughlin got Combat on the radio.

"Cutter *Munro*, 6007," McLaughlin said. "We are en route to you with thirteen survivors on board. That's thirteen POB."

In the back of the Jayhawk, steam was rolling off the fishermen's bodies, water still draining out of their Gumby suits. Nobody said much. Rescue swimmer O'Brien Starr-Hollow was kneeling on the aircraft floor, trying to move around among the pig-piled fishermen. A couple of guys were clearly worse off than the rest.

One was the man flight mechanic Rob DeBolt had had such a hard time prying out of the basket an hour before. The guy was shivering violently. So were several of the others.

As Starr-Hollow assessed each fisherman, he held his hand against the man's head or chest, and tried to look him right in the eyes as he spoke. Starr-Hollow had recently read a book by Jane Goodall, the primate researcher who spent decades studying chimpanzees in the African jungle. The book had impressed upon him how powerful touch and eye contact can be for helping people connect and relax.

All Coast Guard rescue swimmers are trained as emergency medical technicians (EMTs). Part of Starr-Hollow's job was to provide medical care for the survivors—at least until someone more qualified could take over. He knew morale was key to survival. Feeling happier can help someone feel warmer. Starr-Hollow once read that when people lose faith in their ability to survive, their body temperature drops. He knew it was critically important to keep the men awake and alert. On some of the worst-off fishermen, Starr-Hollow used a technique he learned in EMT school, the sternum rub, more or less a medically sanctioned noogie. He made a fist and rubbed hard with his knuckles against the bony part of the central chest. It was painful, but it would keep them awake.

It was about 6:40 A.M. and still dark when the crew on the *Munro*'s bridge spotted the orange-and-white-striped aircraft on the horizon. At the captain's "Tallyho!" the ship's engineers brought the *Munro* down off the turbines and turned her back into the swells. Now they'd be chugging away on their diesels once again, steaming slowly north, away from the wreck. (Originally yelled by hunters to excite their hounds at the sight of a fox, "Tallyho" is also used by pilots and ground controllers to indicate that another aircraft or target is in sight).

The *Munro*'s landing signal officer was outside the hangar door on the flight deck. Outfitted in a neon orange vest and helmet, the LSO held a fluorescent light wand, the same type used by runway technicians at commercial airports. While holding a hover, helicopter pilots can't easily see what's going on beneath them on deck. Instead, they rely on their flight mechanic and the ship's LSO to keep track of the hoists.

The 60 Jayhawk settled into a hover above the flight deck. It had taken the crew forty minutes to fly from the *Warrior* to the *Munro*. The whole time, Evan Holmes remained huddled inside the rescue basket. He had been expecting the chopper to land on the ship. But now he realized that wasn't the way it was gonna go.

Evan wasn't eager to repeat his harrowing ride above the *Warrior*. As flight mechanic DeBolt slid open the aircraft door, Evan looked down toward the stern of the massive ship. *At least it doesn't look like there's much to hit*, he thought as he was lifted out the cabin door and lowered slowly down toward the flight deck. Evan had been parachuting once. This felt similar. The landing was so smooth, he could barely distinguish the moment the basket touched the deck.

BACK AT THE RESCUE SITE, Dolphin aircraft commander TJ Schmitz moved the helo back and to the left so he could see his rescue swimmer, Abe Heller, and the fallen fisherman, who was probably sixty yards away.

Schmitz flashed the helo's landing light, and Heller looked

up, but there was no easy way to communicate why they were trying to get his attention. The pilot panned his light out to where Byron was floating, but in the heavy swells, Heller didn't have an easy line of sight across the water. Besides, he was already working on the next survivor.

Schmitz stared down into the waves. The fisherman was still facedown in the ocean.

The pilot was worried about his flight mechanic. He understood that Musgrave wanted to go back to try to get the man they'd lost. It was an emotional response, and an understandable one. But to Schmitz it didn't make the most sense given the circumstances.

Both he and Greg Gedemer had heard Musgrave grunting and struggling to pull the fisherman into the cabin. Musgrave had seemed to have been at it for at least a minute—an eternity of silence in the middle of a hoist. Musgrave was as strong as an ox, and that's exactly why Schmitz had chosen him that morning. Another ALPAT flight mechanic, Logan Cole, had been scheduled to fly, but Schmitz wanted Musgrave instead. It was unusual for a pilot to hand-pick a mechanic for a specific case, but Schmitz knew Cole was newly qualified. The aircraft commander wanted the strongest, most experienced man available. In his opinion, that man was Al Musgrave.

If Musgrave couldn't have gotten that guy into the helo, well, it was hard to imagine anyone could have.

"We have to move on," Schmitz told his crew. "We have to save the ones we can." He heard a faint "Yes, sir" from the back.

"Are you ready to continue?" Schmitz asked.

Again, a quiet but clear "Yes, sir."

"All right."

"Rescue checklist part two for basket delivery to the survivor," Gedemer announced.

The crew went through the script once again, and then Musgrave had the basket back out the door and moving toward Heller and the fisherman bobbing next to him in the waves.

The basket hit the water and Heller steadied it as Jim climbed in. Thirty seconds later, the engineer tumbled out onto the slippery metal floor of the chopper. He felt like a fish out of water as he tried to move around in his suit. He was shaking badly.

I wouldn't have lasted much longer out there, Jim thought as he pulled himself upright inside the cabin. He could see a few other men pushed up in a ball in the back of the helicopter, some of them all but completely covered up with wool blankets. He scanned the cabin of the tiny copter. Where the hell was Byron?

"Where's the guy that just came up?" he yelled to one of the fishermen next to him.

"There was nobody came up," the man answered. "You were the only one."

Musgrave got Jim Madruga situated and sent the basket back down for Abe Heller. For the first time since they'd arrived at the disaster site, Musgrave pulled Heller all the way into the cabin and slid the door shut.

Schmitz laid out the plan: They'd go back to the group of four who had linked arms in the water, take one, and leave Heller on scene with the rest. With four survivors and two crew in the back, the tiny cabin was already packed full. Leaving the swimmer would make space for one more person.

Heller still had his swimming helmet on, so Musgrave relayed the plan to him. Heller nodded a yes, then grabbed Musgrave's microphone to talk to the pilots up front.

"Leave a raft!" he yelled into the mic.

"We're not gonna leave you without a friggin' raft," Schmitz answered as he brought the helo into a hover over the four men. "There's no way."

Once again, Heller was lowered to the water in the basket. Then Musgrave conned Schmitz about forty yards upwind of the survivors. He pulled out the Dolphin's raft, popped its inflation handle, and kicked it out the open door. The six-man model was the crew's own emergency craft. If the helo crashed, this raft was their best chance at survival, but there was no hesitation over leaving it behind. The raft landed within fifteen feet of the fishermen below. It hit the water upside down, but, as designed, it self-righted when the canopy's support tubes inflated.

Schmitz was impressed with Musgrave's focus. The flight mechanic's conning had been perfect. In these conditions, it would have been so easy to throw the raft too far downwind and for it to blow away from the survivors instead of straight into them. But it was right there.

Down in the waves, Heller had already reached the chain of survivors. He'd sized up their conditions, and pulled the weakest man out of the line. Julio Morales saw the Coast Guard swimmer coming toward him. He felt the man's arms circle his chest. The rescuer was wearing a mask and a snorkel; a small light was attached to his helmet. "You'll be okay. U.S. Coast Guard," the swimmer said.

Julio couldn't believe it. This guy must be crazy to voluntarily jump in the water in these conditions, he thought.

Heller grabbed onto Julio and steered him away from the other men. Heller had the Guatemalan man by the arm when all of a sudden the raft was drifting right toward him. The swimmer hadn't seen it coming.

"Change of plan," the swimmer yelled to Julio.

It would be too difficult for Heller to hold onto the raft while he simultaneously tried to get Julio into the basket. Instead, he loaded Julio straight into the raft, then swam back toward the remaining trio. Heller grabbed the next man he reached. He'd be the one in the basket. The lucky one.

APPROXIMATELY SEVENTY MILES TO THE NORTH, the 60 Jayhawk hovered above the *Munro*. As DeBolt waited for Evan Holmes to climb out of the basket, Starr-Hollow readied the next fisherman. Many of the worst-off men were closest to the cabin door. They'd been the last to be rescued. Now they'd be the first to get out.

As soon as Evan hit the deck, two *Munro* crew members grabbed him steady by the shoulders and told him to stay low. Seconds later, another duo grabbed him and pulled him up. Still in his sopping survival suit, Evan was led across the flight deck and along the outside of the hangar to a door that led to a vestibule, where an EMT team supervised by the *Munro*'s corpsman, "Doc" Chuck Weiss, was waiting.

"We're going to take your suit off and get you out of your wet clothes," Evan was told.

"Everything?" he asked.

There were some females there, too. Great. First I'm in freezing water for several hours and now I gotta get butt naked in front of guys and girls? Evan thought. But he was so happy, he didn't really care.

"Just give me something warm," he said.

The medical crew already had out the trauma shears designed to cut through zippers and other tough material. They quickly stripped off Evan's survival suit and the wet clothes underneath. He was wrapped in warmed blankets that had just been pulled out of the clothes dryer, then walked a level down to the makeshift triage center on the *Munro*'s mess deck.

As the 60 pilots maintained a stable forty-foot hover, the fishermen were loaded into the basket one by one and then lowered to the ship. Each time the Jayhawk's basket left the cabin, there were crew members ready on the flight deck. One of them held a metal "grounding wand" with a clip on one end that attached to the ship. As the basket neared the deck, the crewman would reach out with the wand and touch the metal box, effectively grounding the dangerous static charge that builds up in the basket from the helicopter rotors. With the boat pitching, the *Munro* crew members held the basket and hoist line steady as each survivor reached the deck. They worked systematically, taking each hypothermic fisherman from the basket and handing him off to their crewmates like workers on an assembly line.

As soon as they'd been alerted of the emergency, the *Munro*'s crew had started gathering supplies, especially extra blankets and towels. The dense, gray wool blankets are standard-issue bedding on the 378, and they were loaded into the *Munro*'s eight industrial dryers in armfuls. Two decks up in the kitchen, the crew set the ovens to 160 degrees. They rolled up terrycloth towels and lined them up on industrial-size baking sheets until they came out at about 110 degrees, just cool enough not to scald the skin. They'd be held in the victims' armpits and groins, warming the blood that flows through the axillary and iliac arteries.

When Evan reached the dining room, a *Munro* team was ready with emergency medical treatment. They took his temperature and checked his blood pressure. Now Evan watched from beneath a blanket as each of his coworkers got the same treatment. It was pretty cool, what good care they were getting, Evan thought as he wrapped his numb fingers around a hot drink.

Soon the room was full of wet fishermen being checked on by groups of *Munro* crew. Of the men lowered from the Jayhawk, just three were so weak that they had to be put on stretchers before they were brought down the narrow stairway to the mess deck. One of them was Kenny Smith.

The twenty-two-year-old from Pasco, Washington, couldn't stand on his own when the *Munro* crew pulled him out of the rescue basket. He'd tried to get out but felt paralyzed. Inside the vestibule, the medics stripped off his clothes, secured him onto a padded stretcher, and lowered him carefully down to the next level of the ship.

Kenny was a slight guy, but the maneuver was still tricky. The crew brought him into the dining area, laid him on a table, and started taking his vitals. His core temperature was only 90°F—dangerously low.

Should they move him to the sick bay? the EMTs wondered.

"No. You don't want to jostle the body; it can cause the heart to go into a quiver," Doc Chuck Weiss told the crew. "We shouldn't move him again."

Weiss ordered the crew to grab the thermal recovery capsule, a sleeping bag–like sack designed for rewarming hypothermia victims. The bag is made of a thick, fuzzy wool covered with heavy nylon. Back in Kodiak, just over a week earlier, one of the techs at the medical clinic had offered the rewarming bag to Weiss. It was an extra that had been carried on board one of the helicopters for a while. It had cost close to $3,000.

"Sure, I'll take it," Weiss had said. He didn't imagine he'd ever need it. Now the *Munro*'s Doc wished he had several more of the specialized sacks.

The medical crew was constantly monitoring Kenny's temperature. By the time they had him zipped into the rewarming bag with a dozen of the warmed, rolled towels nestled around his body, the young man's temperature was 91°F. Only his face was exposed and the ice was finally melting from his orange-brown mustache and goatee. The *Munro*'s crew was talking to him, but Kenny didn't seem to be responding. Weiss was worried. The fisherman had a look that the medic thought of as the thousand-yard stare. Occasionally, when they called his name, Kenny would glance over, but mostly his eyes were just wandering. He seemed like he was in a fog, slipping back and forth over the line of consciousness.

DOLPHIN COPILOT GREG GEDEMER was keeping an eye on the fuel. When Abe Heller had been inside the cabin with their four survivors, the helo had exceeded its maximum gross weight: 9,200 pounds. The good news was that the more fuel they burned, the quicker they'd get back within limits. The bad news was that the fuel situation was already tight.

Luckily, the wind had been to their advantage. They'd had a tailwind on the flight to the scene and they'd had 30 knots off the nose for most of the hoists. The wind made it easier to hold a hover at a higher weight. Without it, the tiny aircraft may have had to jettison fuel to keep picking up fishermen. Now it was looking like they'd need every drop available.

Right before takeoff, the pilots had programmed the aircraft's computer to calculate a return trip to the same position. It had seemed like a conservative approach at the time. Once they were up, the *Munro* would continue toward the sinking site, perhaps again at close to 30 knots. The Dolphin crew was under the impression that the 60 Jayhawk was dropping their survivors onto the *Alaska Warrior*—or maybe bringing them to Dutch Harbor. Long before they'd reached the rescue scene, the crew of the smaller helo knew the Jayhawk crew had changed their plans and were intending to lower their survivors to the *Munro*. Still, the Dolphin rescuers thought, that course of action wouldn't take anything close to the amount of time their own crew would be away from the ship.

As Heller worked with the four fishermen below in the waves, Gedemer got the Hercules C-130 on comms. The Herc had taken the 65's guard when they first arrived on scene and now the 65 would rely on the C-130 crew to help them double-check the details of their return flight to the *Munro*.

"Where is the ship right now?" Gedemer asked Herc pilot Matt Duben.

Duben fed Gedemer the numbers. It was just after 7:00 A.M. and the *Munro* was still conducting hoisting operations with the 60 Jayhawk in nearly the same position the Dolphin launched from about seventy miles to the northwest. If the larger helicopter also needed to refuel, the *Munro* would continue moving steadily away from the Dolphin's position.

Helicopter pilots generally fly with a set turnaround in mind. They call it their "bingo"—the point at which they have to head back if they want to arrive with enough fuel for a safe landing. A missed bingo can lead to "splash," a crash of the helo into water. When Gedemer programmed the aircraft's computer at takeoff, he'd calculated a bingo with 400 pounds (59 gallons) of fuel remaining, a fairly typical load that meant they'd land with about forty minutes of flying time left. When the Dolphin got on scene, though, Gedemer bet the *Munro* would be closing some of the distance between their launch point and the disaster site, and he reset the computer back to a bingo of 200 pounds. There were so many people in the water that cutting it closer seemed warranted.

Up in the C-130, the crew went over Gedemer's calculations. There was no question that given the *Munro*'s current location, it was past time for the Dolphin to head back to the ship—the only possible landing point in the middle of the Bering Sea.

"Coast Guard 6566, this is rescue 1705," Duben's cooing

Southern accent broke in over the radio. "You should depart scene as soon as possible, over." Duben was alarmed at how little fuel the helo had, but he didn't show it.

The Alabaman's calm, pillow-talk voice was a comfort to the 65 pilots. Still, they knew the situation was growing serious. They'd pushed the limits and now they had to do every single thing right if they wanted to get their survivors—and themselves—safely back to that tiny patch of blacktop on the stern of the *Munro*.

"Roger, 1705," Gedemer answered. "This is our last guy."

Gedemer was relieved when the basket reached the cabin door and Musgrave quickly pulled it in and unloaded the man inside. With six people in the back, the flight mechanic couldn't fit the metal compartment in sideways and still close the aircraft door. Instead, he propped the basket upright in the space that—until a few minutes before—had held the crew's emergency raft. Then he closed up the cabin as Schmitz began to orbit the tiny, six-man raft below.

The pilot wanted to be sure the fishermen and his swimmer were safe. He asked Musgrave to pull the Dolphin's bright orange data marker buoy from the aircraft wall. The buoyant, arrow-shaped device sends a satellite signal that allows rescuers to relocate a point in the ocean and track the drifting of debris in the water.

Musgrave dropped the marker out the door. The Jayhawk would get back to the scene first. When they did, they would be able to pinpoint the buoy and hopefully find the raft—with four men safe inside—nearby.

BACK UP ABOVE THE *MUNRO*'S FLIGHT DECK, Brian McLaughlin was also updating his fuel calculations. If they didn't refuel now, the aircraft commander realized, they'd have only ten minutes on-scene before they had to turn around again. By the time the last couple of men were being lowered to the ship, the crew knew they'd need to HIFR—now.

Helicopter In-Flight Refueling is a maneuver that all Coast Guard helicopter pilots are well aware of. McLaughlin knew the procedures—even though his training with the *Munro* a couple of days before was the only time in his career he'd actually practiced the refueling technique.

It was 7:10 A.M. when the last hoist to the flight deck was complete. DeBolt pulled the basket back up, and the Jayhawk moved off to the side of the ship. The *Munro* kept heading north into the swells. It was dangerous to be up on turbines during the refueling sequence because the turbines' exhaust could create a "burble," a pocket of hot air that might jostle the helicopter out of the sky. The ship would have to continue at the slow, stable speed they'd used during the hoist—plowing into the waves and moving away from the rescue site.

Down on the flight deck, the LSO signaled to the pilots to keep off to the side of the ship while the HIFR rig was prepared. Like the flight crews, the LSO wore night vision goggles, which lit up the deck in a neon green light.

Meanwhile, the blue-clad tie-downs began to lay out the fuel hose in a big *S* pattern. They called it "faking out" the hose. One of the crew attached a special nozzle that weighed nearly 100 pounds to the hose's end. At the LSO's signal, the 60 moved back over the flight deck and lowered its hoist line with the talon hook on the end, and the "blueberries" attached the nozzle on the fuel hose to the hook.

Mechanic Rob DeBolt had removed a couple of panels to access the interior fuel port, used exclusively for in-flight refueling. The Jayhawk was about thirty feet above the level of the flight deck as DeBolt hoisted the nozzle, with hose attached, all the way up. The hose hung from the hook outside the cabin door as Bonn slid the helo off the side of the ship. Keeping clear of the deck would allow the pilot a better view of what was going on below. The position would also be safer if the fueling hose fell from the aircraft in an emergency. Any spilled gas would fall straight into the ocean rather than onto the deck of the moving ship.

The higher the Jayhawk hovered, the harder the *Munro*'s pumps had to work against gravity to get the fuel up into the aircraft. Forty feet above the surface, though, was about as low as they could be and still stay safely clear of the breaking waves. DeBolt pulled the hose inside the cabin and inserted the heavy metal nozzle into the fuel port.

Like the pilots, DeBolt had limited experience with the HIFR maneuver, and all the experience he did have was in daylight. It was a tricky operation no matter what; attempting it in the dark made it orders of magnitude more difficult. The procedure was sort of like trying to gas up a car with both the vehicle and the gas station moving—along a rutted dirt road. The flight mech kept an eye on the moving ship as he inserted the nozzle into the side of the helicopter. The numbers on the Jayhawk's fuel gauge began to rise steadily. The crew wanted a full tank: 5,500 pounds of fuel.

Meanwhile, Bonn was piloting by the commands of the LSO officer below on the *Munro*'s deck, which was lined with little light-reflecting chiplets that helped Bonn identify the contours of the ship's platform. Every so often, staticky chatter from the 65 Dolphin was audible over the radio.

At 7:28 A.M., just a few minutes after the 60 Jayhawk began in-flight refueling, the crew of the smaller 65 Dolphin reported to the Herc that they had recovered five survivors and left their rescue swimmer—along with their crew life raft—on scene with three more fishermen. A few minutes later, the 65 crew relayed the same message to the *Munro*.

"We have five survivors on board. We left our rescue swimmer on scene," the 60 crew overheard 65 pilot Gedemer report. "We are twenty minutes out and have thirty-six minutes to splash."

Moments later, the *Munro*'s fuel pumps shut down. The 60 was still more than 1,000 pounds shy of a full tank, but there was no time to keep going. Flight mechanic DeBolt removed the nozzle and hooked it back onto the hoist. They repositioned over the deck and lowered the hose back to the ship.

With enough fuel for another three to four hours of flight, the Jayhawk sped south while the massive ship swung back around in the same direction. Down in the engine room, the crew fired up the *Munro*'s turbines once again, hoping to close the gap between their location and the fuel-critical Dolphin.

THROUGH THE SMALL PORTHOLES on the mess deck, some of the crew assigned to medical duty noticed the ship changing its course, turning to the south, the same direction as the crashing waves. A couple minutes later, the massive cutter jolted forward as the engineers brought her up on the birds one more time.

Processor Kenny Smith was in the rewarming capsule and laid out on top of one of the mess deck's tables, which was covered with the heated, wool blankets. The Munro's EMTs were constantly monitoring him. Some of the other *Alaska Ranger* crew members started wandering over as well.

"Hey, bro," one guy said.

"Come on, Kenny, you're gonna be okay, man."

Doc Weiss encouraged the fishermen to keep talking to their friend. The worst thing would be for Kenny to fall asleep and slip away.

Almost all of the rescued men were able to sit up now. Once they had dried off, they had been given "dummy suits," lightweight Tyvek coveralls that are most often used to clothe illegal migrants intercepted in the Caribbean and Gulf of Mexico. Despite their improved appearances, Weiss noticed that most of the fisherman still looked a little stunned. A few were complaining about sore throats. They'd sucked in a lot of salt water, and some of the *Ranger*'s diesel fuel along with it. Some men could still taste it. A few complained of nausea, headaches, or congestion. There had been a few cuts to treat, minor scrapes and abrasions. Overall, though, Weiss thought the men seemed to be doing remarkably well.

Even Kenny—the worst off of all of them—had improved. After ten minutes in the rewarming bag, his core body temperature had risen almost five degrees. He was responding to his friends, smiling, and talking a little. He was going to be okay, Weiss realized with relief. If he had come in even one or two degrees colder, he probably wouldn't have made it.

Within an hour or so, some of the rescued fishermen started leaving the mess deck. The *Munro*'s crew had set up a couple of the ship's TV rooms, the "rec decks" they called them, for the survivors. The fishermen could relax in there, watch a movie, and get their minds off of things.

Meanwhile, Weiss and his team started preparing for the next delivery.

Chapter Twelve

Death at the Extremes

Eric Haynes had been in the *Ranger*'s number three life raft for close to three hours when a bright light illuminated his raft. It was a ship. The *Warrior.*

The trawler drew nearer and within a minute the raft was right up against its starboard side. The *Warrior* was even bigger than the *Ranger,* the black metal hull towering almost two stories up from the water's surface. Eric was on the far side of the raft, watching through the open door on the opposite side. The situation didn't look good. How were they going to get Joshua—the processor whom Eric had found facedown in the water—off the raft? Never mind anybody else.

As the *Warrior* nudged up next to the raft, Eric watched a big hook approach them over the side of the rail. A line was attached and a couple of guys tied it to the raft.

It only took a few seconds before all hell broke loose. The length of line was too short and every time the raft sank into a trough between waves, it was yanked out of the water by the line. Eric could hear the plastic tearing with each jolt.

Also terrifying were several protruding pipes along the side of the ship, the sea side of the discard chutes from the *Warrior*'s factory. Normally, the chutes should be closed when the boat was transiting or at port—but for some reason they were open now. Even when the ship was tied securely at the pier, the sharp protrusions could pop the sturdy buoys that were wedged between the ships to prevent them from banging against one another at dock. They could easily do just as much damage to a life raft, Eric thought. Or to a man.

The *Ranger* crew inside the life raft could see how bad the situation was and began screaming up to the *Warrior*'s rail. "Cut us loose," they yelled. "Cut us loose!"

Eric thought that waiting for the Coast Guard chopper was a better option than trying to board the *Warrior* in these conditions. One of the pipes was big and came to a point just a few feet above the water's surface. He knew that if the raft slammed straight into the pipe, the shelter would deflate and leave ten men in the water, some of them already in a severe state of hypothermia.

Within a couple of minutes, Eric's side of the raft came around against the hull of the ship. The *Warrior*'s crew had hung two Jacob's Ladders over the rail and Eric could see that they were trying to manipulate a life sling up on deck. But the wind made it difficult for the crew on the *Warrior* to get the sling into a controlled position and to lower it to the men in the raft.

One of the Jacob's Ladders kept lurching in front of Eric. One second it was close and the next it jerked ten feet up the side of the boat and landed twenty or thirty feet away. Pretty soon the ladder was in front of Eric again. He could hear the guys up on deck yelling for him to grab it. He reached out and wrapped his neoprene-covered fingers around the rope ladder just in time for it to spring up again with the boat—and slam Eric hard against the rusty hull of the ship.

Eric struggled to secure his feet in the ladder. The crew above was yelling down at him.

"Climb up! Climb up, man!"

He made it only two rungs before the *Warrior* lurched out of the water once again, snapping the ladder hard against the frigid metal. Eric's hands were numb. The ladder slipped from his grip and he hit the water hard. I'm gone, he thought. Even if he avoided being crushed to death by the ship, the *Warrior* probably wouldn't have been able to turn around to get him—not with another nine people off the side of the boat.

But with the next big swell, Eric was thrown back up against the ladder. He wrapped his arms around the rope and hung on with everything he had.

"Pull me up!" he screamed at the rail.

From below, he saw one of the *Warrior*'s crew reach over and hook a line from the ship's small crane onto one of the ladder's orange plastic rungs. Still, the seas kept pounding against the side of the boat. Every time the ship slapped back down into a trough, Eric felt like he was on the verge of being shaken off the ladder. He held on with all his strength as the crane slowly lifted the ladder from the side of the ship a couple of feet at a time.

Finally, Eric was pulled over the rail. The crew tried to help him to his feet, but he couldn't stand on his own. He was lifted up and brought down to the galley, riding piggyback on one of the *Warrior*'s stronger crew members.

The crew helped Eric strip off his survival suit. He was surprised at how wet he was underneath. The *Warrior*'s female observers, Beth and Melissa, helped him into some dry clothes and wrapped him in a blanket. Eric was handed a cup of coffee, but his hands were shaking so badly, he couldn't even lift the mug to his mouth without the hot liquid spilling everywhere.

A couple of minutes later Eric heard a commotion outside. Captain Scott had seen Eric fall in the water. He knew that the *Ranger*'s cook was an exceptionally strong guy; if he couldn't get up that way, it wasn't going to work for anybody else, either. Scott pulled the boat around to try to get the raft in the lee, to protect it from the weather as best he could. He felt like he had a crate of eggs alongside him. If he muscled the boat over too hard, he might easily capsize the raft. Soon, he had the boat at a better angle to the swells and the *Ranger* crewmen were being pulled up out of the raft one after the other. Each man clung to the Jacob's Ladder as the *Warrior* crew used the ship's crane to lift it to deck and then lower it down again.

"One guy is really bad off," Eric warned the observers.

They spread some blankets on a table.

Soon, Joshua Esa was carried into the galley on the ship's metal rescue litter. He looked blue—and just barely conscious. Eric helped Beth and Melissa strip off his suit and cover him up in blankets. The women wrapped their hot potatoes in towels and tucked them around Joshua.

Ed Cook came into the galley as more of the *Ranger*'s crewmen were brought down. Among them was fisheries observer Jay Vallee.

Melissa and Beth embraced Jay.

"Ha! Isn't that wonderful! Oh my God. Get out of those wet clothes," Ed said. "Get out of them!"

"You all right?" Ed asked Jay.

"I'm all right."

"Here's your observer," Ed said to the girls.

* * *

Eric Haynes was feeling better, warmer. He left the galley and headed up to the wheelhouse. The *Ranger*'s cook had known Scott Krey for a couple years. When the ships were tied up in Dutch between trips, Eric sometimes cooked for the *Warrior*'s crew as well. Captain Scott had a surfer-boy appearance that normally fit his laid-back personality. But now he was scanning the seas, his face tense. The captain was intently searching for someone—anyone—else his ship could help. Meanwhile, a few of the *Ranger* crew started crowding into the wheelhouse. "Something happened to the fucking engines and threw everything in reverse," one of them told the *Warrior*'s officers. "The rafts shot way past the bow. Everybody had to fucking bail."

"Once the water hit the generators, the power went out," Eric broke in. He was talking fast. "Then the engines started fluttering, then for some reason it actually went in reverse. You couldn't pull the rafts, they were so tight. . . . We're going over, we're going over, and then two of the rafts, they swung around and came right by the side of the boat."

"I don't know if anybody made it into the other raft," Eric told Captain Scott. "I got to the first one. There were a lot of people in the water. We were trying to get people to swim to us."

Everyone was scanning the waves. The captain had posted some lookouts on the wheelhouse deck and up on the bow.

Finally Scott saw something. "That looks like a raft." He focused the *Warrior*'s spotlight on a yellow disk several hundred yards off in the water.

Jeremy Freitag, the *Ranger*'s steward, was slouched down against the rubber wall. There were twelve of them in the raft. Jeremy felt like they were lost. They hadn't heard anything in a long time. Gwen Rains, the female observer, was extremely upset. Jeremy didn't know her very well; she'd been on the boat for only a few days. He felt bad for her, though. She had some type of beeper, an emergency device that she couldn't get to work. She kept asking the guy next to her if it was broken, but he didn't know, either. Jeremy just bowed his head and tried to block out what was happening. He'd been in the same position for a while when the side of the raft suddenly lit up and a warm, red glow flooded the compartment.

One of the guys on the other side of the raft unzipped the door to the shelter, and there she was—the *Alaska Warrior,* just feet away.

AIRCRAFT COMMANDER TJ SCHMITZ was concerned that he'd overcommitted himself—and his crew. If the *Munro* didn't make up some significant distance, they'd have just minutes to spare before the helicopter ran out of fuel and crashed in the ocean. The last thing Schmitz wanted to do was to ditch in the Bering. In the dark. With five survivors on board.

Normally, an air crew would want 400 pounds of fuel remaining when they landed on a cutter in the middle of the ocean. Maybe 300 would cut it. The absolute minimum—for landing at an airport, in good weather—was 200 pounds of fuel still on board. That was twenty minutes of flying time. At this point, they'd be lucky if they landed with 150 pounds, and there were so many variables between here and there: the weather, the sea state, how quickly the *Munro* could break free of the 60 and get turned around and back on course. It was impossible to predict what could happen with a ship. It could lose radio communications or steering. It could get turned into a big snow squall. On the other hand, an airport runway would always be there.

As they left the rescue scene, Schmitz transferred the controls over to Greg Gedemer. The younger pilot would fly the helo while Schmitz kept an eye on all the gauges. Schmitz decided to slow back. Returning to the *Munro* at a lower speed would decrease the aircraft's burn rate. It would also give the cutter more time to make up the distance between them.

Shortly after 8:00 A.M. Schmitz spotted the *Munro* in the distance. He and Gedemer were still flying with their night vision goggles on, struggling to see as they passed in and out of snow squalls. At first, the ship was just a dot in the green haze of the NVGs.

"Okay, there it is! There it is!" Schmitz announced.

A moment later, a snow squall moved through. The pilot got a sinking feeling.

"Oh, man, we're farther away than I thought."

When they were about five miles out Schmitz called, "Tallyho!"

A few seconds later, Schmitz, Gedemer, and Musgrave all heard the reply: "Tallyho!"

Schmitz looked into the distance and watched as the *Munro* slowed and then turned into the seas.

From the bridge to Combat to the open hangar of the flight deck, the atmosphere on the ship was tense. Much of the crew had heard the exchange between the 65 pilots and Combat; they knew that the small helo was dangerously close to "splash." As the 65 approached the *Munro*'s stern, Schmitz took back the controls from Gedemer. It was a straight shot in.

Schmitz might have only one chance to approach the ship and get the aircraft safely down on the flight deck. There'd be no room for error.

He could overhear the orders urging the tie-down crew back out on deck to await the helicopter. Then SAR Operations Specialist Erin Lopez's voice came over the helicopter's radio:

"Rescue 6566, you're out of limits, but you're clear to land," she said.

Schmitz approached the side of the ship.

Slowly, steadily, he slid the red capsule over the flight deck and planted it right down onto the talon grid. He deployed the talon hook to the honeycombed surface, and the squad of four blue-suited tie-downs scurried out to secure the helicopter to the deck.

Schmitz was out of the helo even before the rotors stopped spinning. He handed control of the shutdown to Gedemer and headed straight to Combat. The dark room smelled like stale coffee. Crowded around a glowing computer bank were Erin Lopez, Chief Luke Cutburth, and Ops Boss Jimmy Terrell. Soon Captain Lloyd came down from the bridge.

"Hell of a job, Lopez!" Schmitz said to the OS.

"You too," Lopez replied. "Thanks for not dying!"

"You scared the shit out of me," Schmitz said. "I think I'll have to check my drawers."

The SAR officer laughed. "Me, too."

"Has anyone heard from Abe yet?" Schmitz asked the assembled crew. They hadn't. Schmitz was concerned. He knew the Hercules C-130 was out over the rescue scene and his rescue swimmer, Abe Heller, should be able to communicate with the plane from his handheld radio. The Herc crew could then pass on the message to the *Munro*'s Combat room.

Erin Lopez had passed on the location of the Coast Guard life raft to the Jayhawk. The larger aircraft, she told Schmitz, was on its way to the coordinates that the 65 had recorded.

"Got a minute, Captain?" Schmitz asked Craig Lloyd. The two men moved over to a back corner.

"We lost one of the survivors," the pilot told the captain. "One guy fell out of the basket, from near the rail."

The captain was silent for a couple moments. He looked at Schmitz. "Okay," he said. "You did the best you could. You had to keep going."

Abe Heller was the only rescue swimmer deployed on the *Munro,* but Schmitz wanted to go back out, even without a swimmer. The *Munro*'s crew would be refueling the helo right now.

"Do you need anything?" the captain asked.

"We're good," Schmitz said. "We're ready."

Up on deck, the crew decided to use the already-set-up HIFR rig to refuel the aircraft. It was an unconventional move but would save time. Each time an aircraft fueled from the ship, two crew members were assigned to check the quality of the fuel. They had to test it for contamination both before and after each refueling to be sure it met "clear and bright" standards. Those two crew members were dressed in purple, rather than the standard blue coveralls worn by the tie-downs. During the entire refueling operation, one *Munro* fireman, dressed in a full-body flame-retardant suit, would remain on deck holding a fire extinguisher. The bulky suit is made of a crinkly silver material that from a distance resembles tin foil. Some of the ship's officers referred to the trio as "two grapes and a baked potato."

Schmitz headed back out to the flight deck. They'd take off again as soon as they had a full tank. The aircraft commander knew that without a swimmer, it would be near impossible to recover anyone who was severely hypothermic. But he figured that if his crew got back to the scene soon enough, they could recover Heller and keep going. And even if they couldn't recover Heller, they could still lower the basket. A lucid survivor might be able to climb in on his own. More likely, though, they'd just be marking locations, helping to find people in the water and waiting for someone else to come to pull them out.

* * *

SEVENTY MILES TO THE SOUTHEAST, the Jacob's Ladders were still hung over the side of the *Alaska Warrior*. In the moments after Steward Jeremy Freitag's life raft reached the ship, a couple of guys had flung themselves at one of the ladders and managed to scramble up.

The recovery process had been improved since the first raft was recovered about forty minutes earlier. The *Warrior*'s deck crew sent down a line on a hook, and the raft was tied off to the ship. Then they pulled up the hook and lowered it again, this time with the ship's life sling attached.

The life-saving device looked like a noose, made of wire coated with plastic. The ship carried it in case of a man overboard accident. In theory, a person in the water would pull the loop down over his upper body and hang his arms over the coated wire. When he was raised, the noose would pull tight, preventing him from falling through.

From inside the raft, Gwen Rains had watched the couple of men struggle up the ladder at the side of the boat. She felt like her range of motion was extremely limited. With her suit on, there was no way she'd be able to do what the men had done, to grab onto the Jacob's Ladder and pull herself up by brute strength.

Gwen heard one of the Japanese technicians yelling at her from the other side of the raft. She crawled over to him. The Japanese man had the life sling in his hands. He motioned for Gwen to raise her arms.

Gwen would be the first one in her raft to be pulled up in the sling.

Please don't drop me, she thought, as she flew up over the rail and landed smoothly on deck. She knew she was almost safe—almost home.

After seeing how efficient the life sling was, no one wanted to throw themselves at the side of a moving ship and potentially end up submerged in the water again.

Jeremy watched the people in his raft being pulled up, one after the other. It was quick. He was the second to last to go.

The life sling felt sturdy around him. He positioned himself right by the door. And then, up he went. He was flying. He felt great; it was almost fun. In seconds, Jeremy had two feet planted on the deck of the *Warrior.* It was the happiest moment of his life.

Warrior Chief Engineer Ed Cook had been running back and forth between the wheelhouse and the deck of the ship as the *Ranger* crewmen were pulled up one after the other. He kept hoping to see his brother emerge from one of the rafts. But Dan Cook wasn't in either one.

Ed went into the galley to check on the rescued fishermen.

"Hey, Chief, you made it!" one of the *Ranger*'s crew yelled when he saw the *Warrior*'s engineer walk in.

"No, man, that's the short one," someone corrected before Ed could say anything.

Ed was nervous. Then again this was less than half of the crew. There was still another raft, wasn't there? And the Coast Guard had been out there for hours now. Maybe his brother was with the Coasties. Possibly he was even one of the guys stuffed inside that first helicopter, the one that had been hovering over their ship a couple hours earlier. Danny might be warming up on the Coast Guard cutter right at this moment.

Ed watched as the fisheries observers, Beth and Melissa, worked on the *Ranger* fishermen. They were taking off their suits and clothes. Those two girls were like angels of mercy, Ed thought. They were drying the guys off and massaging them to try to get the blood flowing again. Just talking to them, trying to help them feel better—adding some calm to the chaos.

Beth stood over one of the rescued men. She had beautiful long hair that fell in auburn waves down toward the man's chest. He was naked on the tabletop and in good enough shape to still be thinking like a guy who's been on a ship for a few months.

"Hey, aren't you supposed to—" he started.

She looked down at him. "What? Make your dreams come true?" Beth smiled. "I'm not gonna do it!"

Everyone laughed. Most of the guys knew that when a man falls overboard, the recommended treatment when he is pulled up is to strip him down and for someone else to get naked too, and lay against him, maybe inside a sleeping bag or under a pile of blankets. Body heat is said to rewarm more efficiently than anything else. Naturally, having a woman in the equation made the whole process seem a lot more appealing.

Up in the wheelhouse, Captain Scott Krey kept scanning the waves. The *Warrior* had saved twenty-two people out of the two rafts, including both of the *Ranger*'s government observers. Between the two helicopters, the Coasties had reportedly rescued another eighteen guys. That left seven men still unaccounted for. Most of the *Ranger*'s crew had abandoned ship around 4:15 A.M. It was now after 8:00 A.M. Everyone else out there had been in the water for a long time—and would almost certainly be in worse shape than any of those rescued so far.

BACK INSIDE THE DOLPHIN HELICOPTER'S six-man life raft, rescue swimmer Abe Heller was focusing on keeping his survivors awake. The three men he'd gotten into the small raft had most likely been in the water for more than three hours. The last man—

Samasoni Fa'aulu, a big Samoan guy everyone called Sam—had been particularly difficult to get into the shelter. He was weak from cold, just dead weight in the water. It took Heller almost ten minutes to push him up and into the life raft. Processor Julio Morales and Assistant Cook Mark Hagerman were already inside, trying to help haul up Sam. Heller went under and tried pushing up on the bulk of the man's survival suit with the top of his head. Finally, the three of them wrestled the Samoan into the raft.

After the 65 Dolphin flew away, Heller had pulled out his EPIRB. He turned it on and attached it to a Velcro patch on the top of his helmet. He could hear the Herc above him and glimpsed the plane now and then through the snow squalls.

In the distance, Heller could see the light of a ship. He hopped back out into the water and pulled out his radio. From his seated position inside the raft, it was difficult to access his gear. And though the craft was designed for six, it was crowded with just four.

"*Alaska Warrior.* I'm a Coast Guard rescue swimmer," Heller said into his waterproof VHF. "I'm in a raft with three hypothermic survivors."

The reply was broken, but Heller got the gist. The ship was busy hauling people out of the *Ranger*'s life rafts.

They'd get there when they could. It might not be soon.

Heller climbed back inside the shelter. The basket-shaped raft wasn't inflated that well, and some water had accumulated on its floor. It seemed sturdy enough, though. The biggest challenge was keeping the fishermen from drifting off to sleep. Sam was in the worst shape.

Julio was also trying to talk to Sam and keep him alert. The two men had been together at the Grand Aleutian hotel before boarding the *Ranger* for the first time just three weeks before.

They'd been sleeping in the same bunk room since then, eating meals together.

For a while, Heller made an effort to get the fishermen talking, to keep their spirits up. He offered the men a candy bar he had stuck in a pocket of his dry suit and tried to give them some good news: "I know another helicopter got a lot of you guys. A bunch of people are okay on the Coast Guard ship."

But the men responded to most of Heller's questions with one-word answers. Every once in a while he shone his flashlight into Sam's face; he had to do something to keep the guy awake. The fishermen didn't want to talk, and Heller wasn't really in the mood for chatting, either. He was exhausted and also feeling a little sick. He'd been slapped around by the waves in the course of the rescue and had swallowed a lot of seawater.

Still, the fishermen were obviously in much rougher shape. *Man, I never want to be on a sinking boat,* Heller thought to himself. The swimmer was concerned about preserving the battery power on his handheld radio. He could still hear the C-130 overhead. He knew his helo crew would have taken his position when they left the scene. What was the point of getting out the radio again? He'd just be telling those guys what they already knew—that he was here with some fishermen, and that they needed to be rescued.

The sky was just beginning to brighten when Heller heard the rotors.

He thought again about pulling out his radio. *I can't tell them anything they're not going to figure out by looking at us,* he told himself. He jumped back into the water and told Julio and Mark to help him pull Sam back into the ocean. By the time the 60 Jayhawk was overhead, Heller had swum with Sam away from the raft. As soon as the helo was in a hover, Heller gave his fellow Coasties the thumbs-up.

They were ready for pickup.

DAYLIGHT WAS BREAKING, AND THE SEAS began to calm. From the *Warrior*'s wheelhouse, Captain Scott spotted something in the water. A splotch of red in the waves—a survival suit. As the ship got closer, Scott could tell there was someone inside. The captain aimed for the suit and slowed his speed as the ship drew near.

Ed Cook went down to the rail. It was a man in the water, all right, floating on his back. His face seemed to be several inches beneath the surface. The man's eyes were open. His mouth, too.

It was Pete Jacobsen. Captain Pete.

Ed knew the captain was a longtimer with the Fishing Company of Alaska. He was close to retirement age now, maybe sixty-five years old. Ed had never heard anyone have a harsh word for the man. Jesus, he thought. Pete, man, just hold on.

It took a dozen men, Ed among them, to pull the *Ranger*'s captain on board the *Warrior.* As the body came up alongside the ship, the men threw a grappling hook to draw it closer to the hull. Even from the wheelhouse, Scott could tell that Pete's survival suit was full of water. The captain was a tiny man, but even with a bunch of guys hauling, they couldn't get him up. It was taking forever. Finally, Ed grabbed a knife and slit the legs of Pete's suit. The water poured out. They got him on board, and then the captain was carried to the *Warrior*'s galley.

Eric Haynes heard a commotion outside. People were arguing about whether to bring the man just recovered from the water down to the factory or into the galley. The crew was getting upset.

"Bring him in here," Eric ordered. "Bring him into the galley."

The crew did as Eric said and laid the *Ranger*'s captain out on a galley table in his mutilated survival suit.

Eric turned Pete's head.

He was gone. There was no doubt in Eric's mind. He could tell by Pete's eyes: They were blank, with no clarity. His suit looked way too big. The captain was sopping wet.

They'd move the body down to the factory. It was too upsetting to see Captain Pete there in the galley, where some of the fishermen were still struggling to warm up.

As Pete Jacobsen was being carried downstairs, Ed heard someone yell that there was another person off the starboard bow.

The *Warrior*'s chief engineer approached the bow rail. About thirty yards off the starboard, Ed saw the body.

"God man, this is a big guy," he said under his breath. "God. Man."

Ed was breathing hard. He wasn't sure if he wanted to see what was coming. He left the deck and climbed up to the wheelhouse as the man was pulled aboard. After a few minutes he got up his courage and headed back downstairs to the galley.

He got there just as the men were carrying the large body inside.

"Don't look, Chief. It's your brother," a crewman said.

Ed walked toward the body.

"I have to look," he told his crewmates.

The man was still in a jumbo-size survival suit. The crew were holding him like pallbearers. Ed moved toward them and pulled back the flap over the man's face. He looked like he was just asleep, like maybe you could shake him and wake him up.

"Yeah," Ed said, as a dozen of the crew stood around the body. Ed's eyes grew wet. His voice cracked as he said aloud what everyone already knew: "That's my brother. That's my little brother Danny."

* * *

Quickly, the crew made a decision: They'd bring Dan Cook down to the factory, too. The *Ranger*'s engineer was six foot two and close to 280 pounds. In the suit, he was too big to get down through the stairwell. They strapped him to the litter that they kept on board in case a crew member had to be medevaced off the ship. Then they opened a hatch, lowered Dan Cook through it, and laid him on the metal packing table in the factory, near Captain Pete.

Eric had climbed over Pete and turned him over to try to get some of the water out. He told the men to cut off their officers' clothes, and sent people for knives and blankets. Eric noticed a sizable wound on Pete's arm, and sent a crewman to find some first aid materials to wrap it up. There was no blood, just exposed muscle. Eric worried that if Pete's body warmed up, the cut might start bleeding. Better to take care of it now.

Within a few minutes, the crew had the two lifeless men out of their clothes and wrapped in blankets. They started massaging their limbs, trying to warm them up. Eric ran up to the wheelhouse, looking for a portable defibrillator.

There wasn't one.

"Should we try CPR?" Eric asked Captain Scott.

The two men decided it was worth a try.

Eric rushed to recruit a group of people to work on the lifeless men. There was some reluctance—it seemed pretty clear to everyone that the officers were dead.

"We have to do this!" Eric told the crew.

Observer Beth Dubofsky began CPR on Captain Pete. It seemed like she was one of the only people on board who had recent CPR certification. She didn't have a mouth guard and the captain looked far gone. His arm was broken and Beth could see some bone sticking out. There were no signs of life, but Beth kept at it.

Meanwhile, Eric was repeating compressions into Dan's chest when a third body was carried into the factory. There was vomit on the man's face. His brown eyes were glazed over. Like with the other fishermen, the crew removed his clothes and wrapped him in blankets. David Hull recognized the man. It was one of the new guys, Byron Carrillo.

David wiped away the vomit and began CPR on Byron. Only a couple days earlier, David had noticed that the greenhorn, while a hard worker, was having trouble distinguishing between different types of fish on the processing table. David had pulled Byron aside and offered him some pointers. Later, Byron put his arm around David up in the galley. "This guy is my teacher," he had said. Jesus, David thought, that was only twelve hours ago. Now he was standing over Byron, pushing water out of the poor guy's body.

"Pump down on his chest. Hard! Fifteen times—and don't worry if you feel a rib pop or crack," Eric instructed David. Beth was standing nearby. The current recommendation was thirty compressions, the fisheries observer told them.

The men decided to go with Beth's number. David did thirty reps, then mouth to mouth. Eric was doing the same on Dan Cook, with a *Warrior* crewman doing the breathing. Over and over. They just kept working.

The men were telling him to stop, but David Hull didn't want to give up on Byron. It just didn't seem fair that he would be the one who didn't make it. For Christ's sake, he'd only been in Alaska for one damn week.

David had been at it for probably fifteen minutes when the *Warrior*'s first mate, Ray Falante, ordered him to stop. There was no hope, Ray told David.

Byron was dead.

Chapter Thirteen

The Final Search

Julio Morales had never been inside a helicopter before. From his seat on the cold metal floor, he couldn't see much of anything outside the cabin. His sweatpants and sweatshirt were wet underneath his torn survival suit. And he couldn't stop shaking. His whole body was so numb he was sure he wouldn't feel the knife if someone stabbed him in the leg.

Julio had been the last of the three in the raft to be pulled up to the Jayhawk. The rescue swimmer sent Sam up first. Mark went next. Then Julio had climbed into the basket, making sure his arms and head stayed inside, like the swimmer instructed. The ride had been fast. Julio couldn't wait to get out of the basket. As soon as it was pulled into the cabin and tilted on its side, he crawled right out and found a spot next to Sam.

Once all of them were inside the chopper, the big Samoan

man peeled off his survival suit's red hood, freeing a huge, springy afro. One of the pilots had to tell Sam to move; his hair was obstructing their view. It was kind of funny, actually. For the first time, Julio felt like everything was going to be okay.

The pilots had spotted another strobe light as they were hovering over the Dolphin's life raft. Once Abe Heller was safely in the cabin, they taxied straight to the light. The person was in a survival suit and floating faceup in the water, the attached strobe flickering on and off amid the waves. They settled forty feet overhead. This fisherman, the pilots saw, wasn't waving up at them like most of the others had.

Julio couldn't hear what the pilots were saying over the roar of the rotors, but he could tell they weren't leaving yet. They were still looking. Julio could see flashes of ocean outside the open aircraft door. It was still dark, but he knew morning would be there soon. Julio watched as Jayhawk rescue swimmer O'Brien Starr-Hollow clipped his harness into the end of the hoist cable. What's he doing? Julio wondered. The rescuer wasn't bringing the basket with him.

The helo crew wanted to move quickly, so they used the rescue strop instead of the basket. As Bonn held the aircraft steady over the fisherman, McLaughlin once again called out the size and frequency of the swells for flight mechanic Rob DeBolt. They were still dealing with huge waves that made it difficult to keep a constant eye on objects in the ocean beneath them.

As soon as Starr-Hollow swung out of the helo, McLaughlin glanced over his left shoulder. The *Alaska Warrior* was headed right at them—and bearing down fast.

The big black boat was shockingly close, its bow pitching in the seas. The top of the *Warrior*'s massive gantry was at about the same height as the aircraft.

Jesus, McLaughlin, thought. They don't see us.

The Jayhawk was sixty-five feet from head to tail—almost a third the length of the ship. Still, in these weather conditions, the helo might be hard to spot. And, most likely, everyone on that boat had their eyes glued to the surface of the water. They would be scanning the swells for bodies—just like the aircrew had been doing minutes before. No one was standing watch for obstacles in the air.

McLaughlin grabbed the radio.

"*Alaska Warrior!* You're headed right for us!" he yelled.

With the swimmer already out the cabin door and a man in the water below, it wasn't an easy option to just fly the aircraft out of the way.

McLaughlin kept watching the ship, waiting for it to turn. Within a couple of long seconds, it swerved out of the way of the hovering helicopter.

The pilot exhaled. Thank God, he thought. A massive ship like the *Warrior* could have knocked his aircraft right out of the sky.

IN THE BACK OF THE CABIN, Julio was oblivious to the near collision with the *Warrior*. His eyes were on the flight mechanic kneeling at the open door. Just a couple minutes after the swimmer descended out of sight, Julio saw the top of his yellow helmet rise back above the floor of the chopper. He had a man secured in a harness in front of him. It took only a few seconds before the swimmer was back in the cabin, but there was a struggle to pull the fisherman fully inside. DeBolt, Starr-Hollow, and Heller were all at the open door, trying to maneuver the guy into the helicopter. The man was lifeless, his legs hanging down below the floor of the helo. Finally, the swimmer reached out with a knife, and made two big slits through the shins of the man's survival suit to drain out the water.

Julio stared as the large man was pulled through the opening and laid down on the slick, wet floor of the cabin. There were ice crystals on his face. Oh shit, Julio thought. He looked at the frozen mustache, the glazed-over eyes. The man had foam at his mouth. His skin looked blue. He was dead, Julio knew it.

It was the first mate, David Silveira. Julio and the other fishermen knew him as "Captain David."

The three rescued crewmen were quiet. They looked at the body.

Julio thought about his conversations with the mate in his first weeks on the ship. He had told the older man about his experience as a marine electrician back in Long Beach, California. David was from California, too—San Diego—and also from a big Catholic family. He had been encouraging and had even offered to help Julio find a full-time job as an electrician in Dutch Harbor.

Julio had liked the idea of it. So far, he loved Alaska. Maybe he could make a life for himself on the island. David had been kind to him. He was a boss on the boat, but he'd acted like he really cared about a new guy like Julio. It was terrible, seeing him like this.

Julio was relieved when one of the Coast Guard crew pulled a blanket over David's face.

UP FRONT, THE PILOTS WERE STILL STUDYING the waves below for signs of life. By their count, three people were still missing. They knew the *Warrior* had already recovered twenty-two people from two life rafts. The pilots saw one of the tented rafts bobbing in the waves. They swooped down to thirty feet to get a good look inside the open door, but the raft was empty. Soon after they found two more. Same thing. There were no more lights in the sea. They looked for red survival suits. Nothing.

The big guy with the afro, Sam, was still in pretty bad shape. Abe Heller had informed the Jayhawk crew that another one of the rescued men, Mark Hagerman, had diabetes. Heller wasn't doing too well himself. The Jayhawk crew could tell the Dolphin's swimmer would be fine, but he was obviously cold and had been puking in the back of the helo. McLaughlin and Bonn decided it was best to get everyone back to the ship as soon as possible.

Daylight broke at 9:07 A.M. But as the orange-and-white-striped chopper skimmed over the Bering Sea, the horizon brightened only from black to gray. There was no sun in sight. The men took off their night vision goggles and squinted into the continuing snow squalls as the helo hurtled north toward the Coast Guard cutter.

DOWN ON THE *MUNRO*'S MESS DECK, "Doc" Chuck Weiss's crew was running on adrenaline. It was almost 9:30 A.M. Most of the Coasties had been up since shortly after 3:00 A.M. They hadn't felt hungry as they waited the long hours for the first survivors, but Weiss had encouraged them to eat. Now they were thankful for the eggs, toast, cereal, and fruit they'd shoveled in hours before.

Finally, they got the word: The 60 Jayhawk was en route to the ship with four survivors on board. One was in critical condition. The smaller 65 Dolphin had just taken off from the flight deck a few minutes before—without a swimmer. The deck was clear for the 60 to lower the new survivors. As the larger helo came into a hover above the flight deck for a second time, Weiss's crew was ready.

Meanwhile, deep inside the ship, the first fishermen to be rescued had settled into one of the *Munro*'s TV lounges. The rec deck was designed like a mini movie theater, with a dozen cushioned, reclining seats arranged on tiers like in a high-end cinema. In front was a flat-screen TV. A couple of the men had picked *The Guardian* from the ship's collection of DVDs. The 2006 film stars Ashton Kutcher and Kevin Costner as Coast Guard rescue swimmers. A few guys had been talking about the movie as they stood on the deck of the sinking *Ranger* just hours before. Now it was almost like they'd lived it. Of course there'd been a copy on board the *Munro*.

The TVs in the *Munro*'s lounges could also be set to the real-time video camera on the flight deck. At 9:40 A.M., three hours after the first and largest group of *Ranger* crewmen had been lowered to the cutter, the rescued fishermen heard the piped announcement that the 60 Jayhawk was about to lower more survivors to the ship. They turned the station to the grainy black-and-white image of the *Munro*'s stern. The rescued fishermen clapped when they saw Sam being led across the ship's deck. There was no mistaking that giant afro.

Julio Morales landed on the flight deck in the basket next and was immediately surrounded. He stood up, and the water inside his oversized suit drained to his feet. The *Munro* crew carried him across the deck, the waterlogged legs of his Gumby suit dragging along behind.

Julio was shivering uncontrollably. He was still being stripped out of his suit and wrapped with hot towels as David Silveira was lowered from the open door of the Jayhawk. The mate was delivered in a seated position, his upper body slumped over his legs inside the basket. Several of the *Munro*'s crew struggled to carry Silveira to the starboard side of the flight deck, where he was transferred to a litter, then brought into the vestibule at the rear of the hangar, where Weiss and a team of EMTs were waiting.

The *Munro*'s crew didn't know what kind of emergency care Silveira had received on the helo, or exactly how long he'd been out of the water. They'd just been told that his condition was critical. They laid the litter on the vestibule floor and checked for a pulse. There was none. Water sloshed out onto the floor as they sliced into Silveira's survival suit. Weiss saw sea foam drool from the man's mouth. His eyes were glassy. He looked dead.

Weiss had received his Coast Guard medical training in Petaluma, California, where he was instructed by Dr. Martin Nemiroff—the Coast Guard's Mr. Miyagi of hypothermia and near drowning. Weiss and his classmates had been lectured on the possibilities for resuscitating cold-water drowning victims. Nemiroff had recounted stories of individuals who had been submerged for a full hour in cold, clean water and later revived without brain damage. "Nobody's dead until they're warm and dead," was one of the doctor's favorite sayings. Nemiroff's stories sounded like incredible, one-in-a-thousand-type cases. But the doctor's own research showed that they weren't as miraculous as they might seem. Over the years, he'd catalogued more than two thousand cold-water deaths or near deaths—many of them from his time with the Coast Guard in Alaska. It sometimes takes a full hour of CPR, Nemiroff told the students, but half of the time cold-water drowning victims can be successfully revived.

"Doc" Weiss found Nemiroff's stories powerful—and convincing. Just because the man in front of him now didn't have a pulse or any other signs of life, didn't mean there was no hope, Weiss concluded. His job was to do everything he could to bring the man back.

Weiss brushed away the sea foam from Silveira's face and placed a sanitary mouth guard used for CPR over the mate's lips. He gave Silveira two breaths and then a crew member linked her fingers together, placed her palms flat just below Silveira's sternum, locked her elbows straight, and began the first of a set of thirty compressions. Silveira was a large man with a broad chest. Weiss estimated he weighed close to 200 pounds. The *Munro*'s doc had never needed to give CPR in a real emergency before, and Silveira was much bigger than the dummies used in training classes. Weiss was surprised by how much energy it took. They needed to get a good two-inch compression with each pump and keep the speed up—the recommended rate was one hundred compressions a minute. The effort required was almost like doing thirty high-speed push-ups, then another thirty, and another. Every few minutes, the person on compressions had to swap out.

They went through five cycles of two breaths and thirty pumps. Then they checked again for a pulse. Nothing. A crew member brought the ship's automated external defibrillator, and they placed it on the fisherman's bare chest. The device listens for a pulse that may be too faint for rescuers to detect and administers an electric shock when a sign of life is identified. But there was nothing to detect. "Continue CPR" the recorded voice on the defibrillator instructed. Two breaths. Thirty thrusts. Again and again, Weiss's team kept at it.

AS THE *MUNRO*'S MEDICS ATTENDED TO the new group of fishermen, Dolphin rescue swimmer Abe Heller was on the mess deck warming up with a hot drink. He was feeling better. He'd stopped shivering and had washed the taste of vomit from his mouth. Heller thought that after a few more minutes he'd be good to go out again if needed.

Dolphin pilot TJ Schmitz had a different idea. From the air, the Jayhawk pilots had passed on the news that Schmitz's swimmer, Heller, who'd stayed behind in the raft, was fairly hypothermic. Schmitz thought that maybe his aircraft could come back to the ship and get Jayhawk swimmer O'Brien Starr-Hollow instead. Starr-Hollow was eager to go. There were still people who were unaccounted for in the sea. They could have spent longer searching the scene, the swimmer thought. Plus, it was pretty amazing to get the chance to work out of two different airframes during the same rescue. Starr-Hollow had never heard of it happening before.

After the last fisherman had been loaded out of the helicopter, Starr-Hollow, too, was sent down on the hoist line. The Jayhawk flew off to the side of the ship as the Dolphin came back into view and landed once again on the *Munro*'s flight deck. With the rotors still running, Starr-Hollow ran to the open door and climbed inside, and the Dolphin took off. Then the Jayhawk came back into a hover—and DeBolt sent the cable down for the HIFR hose. This time, they'd get a full load of fuel.

It was just about 10:00 A.M. when the larger helicopter began in-flight refueling. At 10:15 A.M., the Jayhawk crew overheard the *Munro* order the Dolphin back to the ship: The search was over. The *Warrior* had recovered the final three *Alaska Ranger* crew members—all of them deceased. Now all forty-seven people were accounted for.

At 10:18 A.M. the Jayhawk's fuel tanks were full. When the Dolphin arrived back at the ship, a game of musical chairs commenced like no one had ever played before. As McLaughlin and Bonn flew the Jayhawk alongside the cutter, the Dolphin hovered over the flight deck and lowered rescue swimmer O'Brien Starr-Hollow to the ship. It was too dangerous for the Jayhawk crew to hoist their swimmer from the ship with the smaller helicopter perched on the flight deck, and the larger helo didn't want to spend time and fuel waiting for the Dolphin to be shut down and moved into the hangar. Instead, the smaller aircraft remained airborne.

After Starr-Hollow was safely on deck, the 65 crew pulled the little red helicopter away from the boat, and then the 60 Jayhawk came into a hover over the lurching flight deck for a final time. DeBolt lowered the hoist cable, and Starr-Hollow clipped the hook into his harness. They'd go back to St. Paul together, as a team.

NOT LONG AFTER THE FIRST GROUP of fishermen had been lowered to the cutter, there was a piped announcement that the ship was looking for donations for the rescued men. Coasties filed into the mess deck with spare T-shirts, sweatpants, boat shoes and sneakers. The ship's store was opened up to the men. They could grab a hat, a sweatshirt, and—best of all for many—free cigarettes. Some of the *Munro*'s crew gave the fishermen snacks from their own shore-bought stashes. By late morning, most of the first-rescued fishermen were sprawled out deep in the ship, dressed in crew hand-me-downs or brand-new Coast Guard–branded sweats.

As the hours passed, Doc Weiss kept checking in with the fishermen. Kenny Smith was sleeping. It had taken him a couple of hours to be able to move around on his own. Even once he could, he complained of numbness. There were quite a few guys with bruises and some sprains. A few men complained of what sounded to Weiss like swimmer's ear. But overall, the men were doing remarkably well.

Except for the very last fisherman, David Silveira.

They declared him dead at 11:00 A.M.

After several cycles of CPR on the floor of the cold vestibule behind the hangar, they had moved Silveira into the *Munro*'s sick bay. They'd treated all the other fishermen on the mess deck, but Weiss didn't want to upset the *Ranger*'s crew—or the *Munro*'s own young seamen—with the sight of someone who was so far gone. From the sick bay, Weiss called the on-duty doctor at the air station in Kodiak. Keep going, the M.D. had encouraged the *Munro*'s corpsman. It can't hurt. But when Weiss called back close to an hour later, he was told to stop. Silveira's core temperature was at 83°F. They'd been trying to resuscitate him for seventy minutes already. It was over.

Weiss helped close Silveira's body into a green vinyl body bag. The Coasties brought the body to a sheltered spot outside the XO's cabin and laid it down. Then they assigned a crewman to stand watch over the dead officer.

THROUGHOUT THE RESCUE PROCESS, Combat had been reporting the number of fishermen lowered to the *Munro* to District Command in Juneau. The last four men—including David Silveira—brought their recorded total up to twenty-two. Just before 11:00 A.M., District called off the search. All forty-seven people who had been on board the *Alaska Ranger* were accounted for. The cutter *Munro* had reported twenty-two men—thirteen in the first Jayhawk pickup, five from the Dolphin, and four more recovered by the Jayhawk on that helo's return to the rescue scene. Meanwhile, there were twenty-five people, including three deceased, on board the *Warrior*. At 10:57 A.M., District directed the *Munro* to accompany the *Warrior* back to Dutch Harbor, where both ships would transfer their survivors to waiting medical personnel.

Soon after the second life raft was unloaded onto the FCA trawler, the rescued *Ranger* crew were asked to write their names on a sheet of paper, and those names were passed off to District personnel. A few dozen miles to the north, a Coastie on the *Munro* made a list of names of the *Ranger* crew on the cutter. Meanwhile, SAR officers in Juneau had been working to get a full list of the *Alaska Ranger*'s crew from the Fishing Company of Alaska. By noontime, the Juneau Coasties were comparing the lists. They'd recorded that the *Munro* had twenty-two men, but when they looked at the names, there were only twenty-one. Where was a Japanese crew member named Satoshi Konno? His name didn't appear on the list supplied by either ship.

It was just after 1:00 P.M. when District informed the ships that Konno was unaccounted for. There'd been a miscount. Though crew on the *Munro*'s bridge had counted just twelve survivors lowered to the deck in the Jayhawk's first load, Combat recorded the number they'd heard the helo's own crew report: thirteen. It was that number that was passed on to Juneau. Now everybody knew that one man was still missing. The *Munro* was on course toward the *Warrior* when Captain Lloyd ordered his engineers to make best speed back toward the disaster site.

FOURTEEN HUNDRED MILES AWAY, in the windowless control center at Coast Guard headquarters in downtown Juneau, SAR personnel were gathering information on the missing fisherman—his age, height, weight, and the estimated time he'd been in the Bering Sea—and plugging it into their modeling program. Developed by the Coast Guard specifically to find lost mariners, the computer software uses on-scene weather data, including real-time information about sea currents, to calculate the most probable drift pattern of a person lost at sea and recommend search grids to optimize the chances that a man overboard will be found alive.

Given different variables, the personal data on a lost individual are used to calculate likely survival times. Satoshi Konno was a thin man in his fifties. His age was a disadvantage, and so was his build. By far the fisherman's worst enemy, though, was time. At the moment the Coasties realized Konno was still missing, the Japanese man had been in the water for at least eight hours. The sea temperature was 36°F.

The *Munro* made it to the sinking site around 2:00 P.M. with a search plan ready. Crew members were posted at the ship's rails to scan the seas while the massive cutter methodically traced search lines across several square miles of open ocean.

Dolphin pilots TJ Schmitz and Greg Gedemer were eager to provide more eyes on the waves. But the 65 was grounded. Byron Carrillo's fall from the basket had been labeled a Class A Mishap, a designation used for an accident that results either in a death, or in total loss of the aircraft. The service's rules for that classification are strict—neither the aircraft nor the crew could fly again until an investigation was complete.

The Jayhawk crew had been told the search was over and left the *Munro* just before 11:00 A.M. It took them just over an hour to fly back to St. Paul Island. They landed the aircraft at 12:20 P.M. By the time they pulled the helo into the hangar and drove the snowy mile back to the LORAN Station, the Coasties more than a thousand miles away in Juneau knew something was wrong.

In the pilots' lounge on St. Paul, the phone rang once again.

There'd been a mistake in the count. One man was still out there.

McLaughlin and his crew had eight and a half hours of flying time on them. If they'd been the only air crew available, their commander could have given them the thumbs-up to go out again. But another crew was on standby. Pilots Shawn Tripp and Zach Koehler would fly the 60 Jayhawk back to the scene.

ON BOARD THE CUTTER *MUNRO*, the news about the missing man quickly trickled down to the *Ranger* crew. One of the Coasties was making rounds among the rescued fishermen, asking questions. Had anyone seen Satoshi Konno abandon ship? How about in the water? Did anyone remember what he was wearing?

At first, Julio Morales had no idea who the officers were talking about. "Who's that?" he asked one of the other fishermen. "Who's Satoshi Konno?" The name didn't sound familiar at all.

"The fish master. It's the fish master," somebody told him.

Julio had seen the fish master in the wheelhouse before he'd abandoned ship. By that time, most people were already in the ocean. The boat was listing and the water was almost up to the deck surrounding the bridge. While everyone else was running around in their bulky suits trying to figure out how best to get off the ship, the fish master had been sitting quietly inside. He seemed calm and was smoking a cigarette. His suit was unzipped down to his waist. He didn't seem to be in much of a hurry to save his own life.

THE FISHERMEN WERE INVITED TO SEND brief e-mails home. They couldn't make any phone calls, though. The authorities were still confirming the names of the deceased and working to notify the families. Until then, the crew could only communicate with their own family members by e-mails, which were to be reviewed by the Coast Guard.

Ryan Shuck sat down in one of the Coastie's bunk rooms to write to his longtime girlfriend, Kami. "The boat went down early this morning," Ryan typed. "Most of us were rescued by the coast guard they have been great. I'm on a cutter now we're searching for 1 guy still. I'll call when I get to dutch. I'm ok don't worry. Love always ryan."

The *Munro* was still tracing search patterns for Konno on Monday morning when Captain Lloyd gathered the *Alaska Ranger*'s crew together on the mess deck. The captain had a piece of paper in his hands.

"I have the names of the deceased," he said.

Julio's cousin Marco had been in the first load of fishermen rescued by the 60 Jayhawk. When Julio reached the ship, Marco was already warmed up and doing fine. Neither man knew what had happened to their other cousin, Byron.

Julio had a bad feeling. But as the captain read the list out loud, he didn't hear his cousin's name. It was the ship's top officers who had died, Julio realized. The captain, the mate, and the chief engineer.

Byron's okay, Julio thought for a minute. We're all okay.

But there had been a name on the captain's list that he didn't recognize. It sounded like Brian. Julio didn't remember meeting a Brian on the *Ranger.* He asked to see the names.

He took the paper in his hands and there it was: Byron Carrillo.

Julio felt like his ears were ringing. He heard himself screaming. "No. No." The other men were looking at him. Julio was sobbing. Some of the other men had tears in their eyes, too. Julio knew that most of them didn't really know his cousin.

Four days wasn't long enough to know him, but they'd liked him well enough. He was a pretty friendly guy, a good worker. Now he was the new guy who was dead.

THE *ALASKA WARRIOR* WAS HEADED BACK to Dutch Harbor. A couple of the other FCA trawlers were searching for Konno near the sinking site. And Captain Scott knew the Coast Guard was out there, completing calculated search patterns with their plane, helicopter, and cutter all scouring the seas hour after hour.

As the *Warrior* steamed east, Ed Cook went down to the factory, to where the bodies were. The crew had put a white blanket over Captain Pete. Danny was covered with a red one. Ed had watched the CPR attempts on his brother. But there was nothing, no sign of life. Danny had been in the water too long. Ed knew it. His brother's lungs were full of water. So much water. It had poured out of him when the crew pumped again and again at his huge chest.

Ed peeled the cover back from his brother's face. Danny had been facedown in the water when they pulled him out. His eyes had been closed. Even in those first few awful minutes, he'd looked like he might just be asleep. Just sleeping, that's all.

Ed stared at his brother. He looked peaceful. He even had a little smile on his face. His lips were pressed together, closed. That was unusual, Ed thought. His brother was a nonstop talker. Now, for once, he was quiet.

Danny looked enormous under his blanket, especially next to the captain. Ed wrapped his arms around his brother's body. He kissed him on the cheek, and then on the lips.

"Daniel, I love you," he said out loud. "I will miss you every day of my life."

Chapter Fourteen

The Investigation

The sky was black as the *Munro* entered Iliuliuk Bay. The ship steamed for the cluster of lights that marked the village of Unalaska, then swerved to starboard, toward the Coast Guard pier. The *Munro*'s crew had searched for fish master Satoshi Konno for twenty-six hours. There'd been lots of debris—fishing net and buoys, a couple of life rings—and a half-mile-wide oil slick. The rescue swimmer on the new Jayhawk crew from St. Paul had been lowered on the hoist line to puncture and deflate the empty life rafts so they wouldn't offer false hope to the next group of searchers. Like the seamen posted on deck as the *Munro* zigzagged up and down the ocean, the helo crew had seen no trace of the Japanese fisherman.

Erin Lopez was back on duty in Combat, studying the radar for any traffic in the channel and supervising the more junior operations specialists as they plotted the ship's approach into Dutch Harbor. Lopez had taken charge of developing the *Munro*'s search patterns for Konno, working with District Command in Juneau to be sure the ship's plans and those of the Coast Guard aircraft complemented one another. As the ship neared the pier, she went up to the deck so she could watch the fishermen go ashore.

In the hours after the last men were lowered to the ship, Lopez had talked to some of the survivors. The room she shared with a handful of other female crew members was right across from the rec deck where the men were resting. She'd poked her head in there and introduced herself, offering the guys popcorn and PowerBars from her private stash. A few of the survivors said they remembered her voice. They'd heard it coming over the radio in the *Ranger*'s wheelhouse before they abandoned ship.

Now most of the fishermen were on the mess deck, waiting to get off the boat. Out the portholes, they could see that there were a handful of vehicles next to the pier, including an ambulance to transport the body of David Silveira.

As the *Munro*'s deckhands raised the aluminum gangway up to the side of the ship, Lopez moved to the rail. It was after midnight, but Captain Lloyd was there, along with Luke Cutburth and Jimmy Terrell. As the rescued crewmen stepped off the boat, Lopez shook each of their hands. A couple of the guys leaned down to give her a hug. Then she watched them file quietly off the ship and disappear into the cold night.

WITHIN A DAY OF THE SINKING, the Coast Guard and the National Transportation Safety Board (NTSB) convened a joint Marine Board of Investigation into the loss of the *Alaska Ranger*. The Marine Board is the Coast Guard's most formal type of accident analysis, most often reserved for incidents that result in multiple deaths, as well as the total loss of a vessel. It had been seven years since the Coast Guard's last Marine Board, which investigated the disappearance of the *Arctic Rose,* the head-and-gut boat that sank in the Bering Sea in April 2001 and cost the lives of fifteen men.

The *Alaska Ranger*'s Marine Board began questioning witnesses on the morning of Friday, March 28, in a conference room at the Grand Aleutian hotel. The board was headed by Coast Guard Captain Mike Rand, then finishing out his career at Coast Guard headquarters in Washington, D.C. Assisting him were three other Coast Guard officers, one from Anchorage, the other two from the East Coast. The four-man NTSB team was led by a young maritime accident investigator named Liam LaRue. Though the fact-finding phase of the investigation would be a joint effort between the Coast Guard and NTSB, each agency would ultimately prepare its own independent report on the disaster.

Marine accidents represent the smallest slice of the pie of all transportation casualties investigated each year by the NTSB, the federal agency charged with examining the causal factors of airline crashes, train derailments, and highway disasters. LaRue was in a good position to lead the investigation. A 2000 graduate of the Coast Guard Academy, he'd spent two years as a deck watch officer on a cutter that had patrolled in the Bering. He then served for three years as a Coast Guard marine inspector before joining the NTSB. In his time with the agency, LaRue had worked on a number of high-profile incidents, including the capsizing of the passenger vessel *Ethan Allen* in Lake George, New York, in 2005, which resulted in the deaths of twenty-one elderly vacationers, and the collision of the freighter *Cosco Busan* with San Francisco's Bay Bridge in 2007.

"This investigation is intended to determine the cause of the casualty to the extent possible," Captain Rand explained in his opening statement, "and . . . to obtain information for the purpose of preventing or reducing the effects of similar casualties in the future."

The board would look for evidence of any "incompetence, misconduct, unskillfulness, or willful violation of the law" that may have contributed to the accident, Rand said. Though the board would examine the decisions and actions of Coast Guard marine inspectors and rescuers, he identified a single "party of interest" to the investigation: the Fishing Company of Alaska. Before calling the first witness, the Coast Guard captain asked everyone in the conference room to stand and observe a moment of silence for the five men who had lost their lives just days before.

Julio Morales testified on the third day. Along with the other rescued men on the *Munro,* he'd been brought by van back to the Grand Aleutian hotel after arriving in Dutch Harbor. The guys who'd been picked up by the *Warrior* were already partying. There'd been another open tab. But this time, the company's rule against alcohol had been lifted. The men could drink as much as they wanted and they'd each been given a wad of spending cash, courtesy of FCA. Some of the guys said the company was trying to butter them up. Julio had more than a few drinks. He kept seeing Byron's face at night—but with enough beer, he could fall asleep.

Julio had been told he had to stay for the investigation. He missed the funeral. The family was Catholic, so the service was held right away. The day after they got off the Coast Guard ship, Julio had gotten a call from the cops in Dutch Harbor. They wanted him to come to the airport and identify his cousin's body. Julio didn't want to do it and asked Marco to go instead. The police came and picked Marco up, and brought him to the airport, where Byron's body was waiting to be flown to the medical examiner in Anchorage. Julio waited at the hotel.

On the witness stand, Julio told the Marine Board about the

holes in his suit, the rip at the seam that had allowed water to leak in. The investigators asked Julio about his training. Had he ever put on the survival suit before the night of the sinking? Yes, Julio said. He'd practiced getting into the suit soon after he'd first boarded the ship. His cousin Byron had done the same when he joined the crew a few weeks later. Other than that, though, there'd been no training. No one told them how to abandon ship, how to board the life rafts from the boat, or what to do once they were in the water.

The next day, the board called fisheries observer Gwen Rains. They questioned her about the safety checklist she'd filled out when she first boarded the *Ranger.* "Did you find any discrepancies during your inspection?" Gwen was asked. There were several things she felt that the ship needed to address, she responded, but the boat did have the current Coast Guard safety decal required to sail with an observer, and none of the boat's problem areas were on her no-go list.

She gave some examples: "Like, the watertight doors, do they seal properly? No, they don't," Gwen said. "Are there fire extinguishers in every corridor that are in good and serviceable condition? No, they're not. There were several things like that."

Gwen recalled the safety drill carried out less than a week before the sinking. Unlike in other drills she'd witnessed as an observer, the men on the *Ranger* didn't actually put on their survival suits or fire suits, or act out emergency scenarios with the full crew. Instead, they mustered in the wheelhouse, and the emergency squad members reported to the officers where they *would* be in different emergency situations. There was no discussion about what to do in an abandon ship scenario.

"What else is there that we need to know?" Rand asked at the end of Gwen's testimony.

"I feel very strongly that a bad situation was made worse by people not knowing what to do. By people not being trained," she answered.

The Marine Board took seven days of testimony in Dutch Harbor and, two days later, reconvened at the Hilton hotel in downtown Anchorage. Jayhawk pilot Brian McLaughlin and his wife, Amy, had flown from Kodiak, along with three crew members of the 65 Dolphin: pilot TJ Schmitz, rescue swimmer Abe Heller, and flight mechanic Al Musgrave. The primary goal of the investigation was to determine why the *Ranger* sank and why so many people failed to abandon ship safely into life rafts. But the board would also look at the Coast Guard's response: What went well; what went wrong; and how could it have gone better.

It wasn't the first time the Coast Guard rescuers had told their stories. When the *Munro* arrived in Dutch Harbor with the *Ranger* fishermen, the crew of the 65 Dolphin packed their things and left the ship. The next day they were flown to the Coast Guard outpost in Cold Bay, where they were met by a team of experienced Alaskan aviators who would conduct an internal fact-finding mission into how one of the shipwrecked men had fallen from the rescue basket. Each 65 crew member was questioned about Byron Carrillo independently, and then they talked together about how things might have gone differently. Back on the *Munro,* each Coastie had been given drug and alcohol tests, which is standard Coast Guard practice after any major mishap. All the men's tests were negative.

Not long after the men were back home in Kodiak, the air station commander announced a safety stand-down—a meeting where all the rescuers involved in the *Alaska Ranger* case would recount what happened and discuss how and why they made the decisions they did. The gathering was held at the movie theater on base and was open to the entire air station—and to any family members who wanted to attend.

It was crowded as Amy McLaughlin walked into the warm theater. Framed film posters hung on the walls alongside signs for the weekday, kid-friendly screenings where she sometimes saw new releases with her young daughter in tow. Amy was curious to hear more about what had happened on the case. But mostly she just wanted to support Brian. She saw a few of the other wives there, too.

Brian had called Amy from St. Paul that Easter night. They'd had a pretty big case, he told her. He sounded exhausted, too tired to tell the whole story. It wasn't until she saw the news the next day that Amy realized just how huge the rescue had been. She noticed that most of the headlines focused on the five who died. Of course, it was heartbreaking for those families, but the endless spotlight on those who hadn't survived made Amy feel bad for the pilots, and for all the guys on those aircrews. She wished the media would concentrate on the ones who'd been saved. It was hard for Brian, Amy knew, to accept the fact that they'd left people behind. He'd never had to leave anybody before. He seemed concerned that she know his crew had done everything they could out there—and she did.

One by one, the Coast Guard rescuers stood before the familiar faces in the crowded movie house and told their stories. They presented the rescue on a timeline: the 60 guys went first, followed by the 65 guys, then back to the 60 crew. They talked for more than an hour, and then took questions. Amy was interested in hearing what the other pilots asked about the decision making. It was by far the most intense rescue Brian had ever experienced. Hearing her husband describe all the lights in the water, the high seas and winds, and the disorienting snow squalls made her feel uneasy. But mostly she felt proud.

The *Munro* was back on routine patrol in the Bering Sea at the time of the safety stand-down in Kodiak. But the ship's crucial role in the rescue was duly recounted by the aviators, including Brian McLaughlin. He had already offered his appreciation personally. On the morning of Monday, March 24, as the *Munro* was still tracing search patterns across the Bering Sea, Ops Boss Jimmy Terrell had received an e-mail message from the aircraft commander:

> *Jimmy,*
>
> *I wanted to drop you a line to try to express our sincere appreciation for your crew's efforts yesterday during the main portion of the* Alaskan *[sic]* Ranger *case. As we left, we asked your Control watch stander to relay that to your CO, but it's kind of hard to get the point across on a working freq. . . .*
>
> *It is vividly apparent to me and my crew that your crew was rolling as continuously as we were from the 3 o'clock hour, and continued to do so even after we left scene.*
>
> *This case covered literally just about every aspect of CG SAR training that they beat into us as pilots: navigating and operating in poor weather, high seas, hoisting survivors in the water and in rafts, hoisting to a pitching fishing vessel, then to a pitching cutter, HIFR ops, search planning, etc. It was by far the most large-scale CG operation that I have ever been involved with, employing 5 aircraft, 7 crews, good sams, etc., all of which centered about the Cutter* Munro. *If you hadn't been there, I can't imagine what the final outcome would have been.*
>
> *As you continue on your patrol and on this case, as I'm sure you are, please know that outside of the obvious numerous survivors that you are bringing back home, your fellow Coasties are well aware of what you*

put forth to make it happen, and are damn thankful that you were there to do it. Please pass our gratitude and sincere respect to your crew.

Semper Paratus.

Brian

WITHIN A COUPLE OF WEEKS, the cutter was on patrol back up near the ice edge. It was a sunny day with no wind, and the water was dead calm near the retreating ice floe. Captain Lloyd piped an instruction for the crew to report to the flight deck. It was time for a swim call. The ship's bagged survival suits were sorted by size. The crew pulled them on and leaped off the stern deck into the ocean. They bobbed up and down in the water, practicing swimming and linking arms in the flat seas. The officers paid $3 each for the privilege, a collection for the ship's morale fund. For the junior crew, the exercise was free. Chuck Weiss observed from the deck. He was impressed by how well the suits worked, and it made him feel good to see everyone out there for so long.

Weiss had seen a few of the rescued *Ranger* crew members again, in the UniSea Bar in Dutch Harbor. He was accustomed to getting some nasty looks from fishermen at the bars in Dutch. The Coasties didn't always feel welcome in there; they were the cops on the water, after all. But after the rescue, when Weiss walked in with a group from the ship, several fishermen came right up to them.

"Hey, we want to get you guys some beers!"

More people approached the Coasties: "You guys off the *Munro*?"

"Yeah," one of the seamen answered.

"Right on!"

Weiss was slapped on the back. Someone ordered him a drink. Soon, there was a round of shots in front of them.

"These guys right here, they're lifesavers," one fisherman announced to the crowded bar.

IN MID-APRIL, THE COAST GUARD and NTSB Marine Board of Investigation convened in a third location, in a conference room at the Red Lion Hotel in downtown Seattle, just a few blocks from the famous fishmongers at Pike Place Market and less than two miles from the Fishing Company of Alaska's corporate office. One by one, the *Ranger* fishermen described what they thought had contributed to the sinking. The engineers often didn't keep consistent watch, some said. Several more testified that there was regular drinking on the boat, often among the ship's officers. A number of men said that in recent months the *Ranger* had been traveling through ice more forcefully than they had experienced before, on this or other vessels.

"The ice is just like being . . . on a frozen lake," Ryan Shuck explained several days into the Seattle hearings. "You see a few little cracks in it. . . . You couldn't really see a lot of water between one piece and the other. It was pretty dense."

Shuck recounted for the board the arguments he'd witnessed between the fish master and the previous captain, Steve Slotvig, including the fight about the boat's speed traveling through ice. And he was questioned about who he'd seen drinking on the ship.

"Quite honestly," Ryan answered, "I'd say probably 80 percent of the crew, at one point or another."

"Did you use alcohol on board?" Shuck was asked.

"I have."

"Was alcohol allowed on board?" one of the Coast Guard investigators pressed.

"No."

"What was the company's policy?"

"The company's policy is no drugs or alcohol on the vessel."

"But you'd estimate 80 percent of the crew was drinking even though that was the policy?"

"Yeah," Ryan said. "I'd say that's probably a conservative estimate."

Ryan was asked if he'd seen any of the engineers drinking. He testified that the assistant engineer, Rodney Lundy, was frequently intoxicated.

"Did you ever see him actually drinking alcohol or beer?" the Coastie pushed.

"Not on the boat."

"So how did you know that he was frequently intoxicated?"

"I guess just the same way that, if you went down to the bar right now and had a six-pack of beer and came back up, I'd probably know," Ryan said to a few grim smiles. "You can just tell."

Many of the *Ranger* crew, including Ryan, answered the Marine Board's questions with counsel by their side. Even before they arrived back in Dutch Harbor, some of the men had been talking about lawsuits, and soon after Ryan Shuck's girlfriend read his e-mail from the Coast Guard ship, she Googled "maritime lawyer." When the rescued men checked into the Grand Aleutian, they were greeted by messages from lawyers wanting to represent them. By the time the board convened in Seattle three weeks later, the crew had split fairly evenly into two groups: those who wanted to keep fishing for the FCA, and those who were ready to lawyer up.

The company had made settlement offers to the fishermen early on: $25,000 to those who got into a life raft pretty much right away, $35,000 to the men who struggled in the water for a time before finding their way to a raft, and $75,000 for those who were airlifted by the Coast Guard. Larger offers were made to some of the more senior crew members who'd played a key role in evacuating the ship. Factory manager Evan Holmes was one of them.

Like many of the *Ranger* fishermen, Evan had lawyers calling him in the days and weeks after the disaster. He was told that he was hurting some of the more junior guys by settling, that because he was a more experienced crew member, his lawsuit would help theirs. The pressure pissed Evan off—and so did some of the things he'd heard the guys saying. Like that they'd called the ship "*Ranger* Danger," as a few crew members had told reporters. It wasn't the boat that people were talking about when they used that phrase, as Evan remembered it. It was the crew.

Evan took the FCA's money. He just wanted it over with. People died, and no lawsuit was going to change that. Evan wasn't planning to go back to fishing—at least not right away. He wouldn't rule it out for the future, though. He still had buddies up in Dutch. It was decent money, after all, and he'd been good at the job.

SEVERAL DAYS INTO THE MARINE BOARD testimony in Seattle, Coast Guard Commander Chris Woodley was called to the witness stand. The *Ranger* had been one of about sixty head-and-gut boats enrolled in the Alternative Compliance and Safety Agreement (ACSA) and Woodley had been the original mastermind behind that program.

The goal of ACSA had been to take a fleet with proven safety problems—driven home by the loss of the *Arctic Rose* and the *Galaxy,* just a year apart—and increase its safety standards. The fact that most of the boats in the fleet had been making and selling fish products that legally only classed and load-lined vessels were permitted to sell gave the Coast Guard the leverage needed to get the program off the ground. Most ships in the fleet were too old to be touched by the class and load line societies, Woodley explained to the Marine Board, and new fisheries management regulations focused on conservation banned companies from replacing old boats with new ones. The companies' hard-fought fishing quotas were "tied to the steel." Even if a company were willing to scrap an aging vessel and spend $10 million or $15 million on a new ship, they'd just end up with a beautiful new boat they couldn't fish.

"We had several options in front of us," Woodley testified. "Our first option is we could tell them, you guys can't make these products anymore. . . . They could simply operate as fishing vessels, engage in simple head-and-gut processing operations, and there would be no increased safety standard for about fifteen hundred to sixteen hundred people that work on board those boats. That didn't really seem to be a viable option. We're about improving safety, not the status quo."

Instead, Woodley explained, they came up with a new program that would bring the boats up to an "equivalent" level of safety as to what the class and load line societies require. Going into the program, the major concerns were about vessel stability, watertight integrity, and the degree to which crew could safely get off the boat in an emergency (an issue that is not addressed by class and load line societies).

"Did [the ship owners] ever express their concern over the workload that it was going to take for them to come into compliance with the program?" a Marine Board member asked.

"Absolutely," Woodley answered. "This is a heavy lift for this fleet. We've estimated—this is a rough estimate—about forty million dollars going into these boats. I think, quite frankly, a lot of owners were very surprised at the condition of their vessels once they started dry-docking them and you had a marine inspector—a Coast Guard marine inspector—climbing the tanks and finding a lot of damage."

The board had already questioned FCA Operations Manager Bill McGill about the FCA's experience with ACSA. The tall, gray-haired captain had sailed as master of the *Alaska Ranger* for more than a decade before advancing to a desk job, and he acted as the spokesperson for the FCA throughout the Marine Board process. (Company owner Karena Adler was never called to the witness stand.) McGill told the Marine Board that it would have made more financial sense for the FCA to stop making the ancillary products that would be banned if they didn't comply with ACSA, than to comply and invest in the significant upgrades the safety program required. Still, the FCA had enrolled their seven vessels and had spent several million dollars on the program to date.

"We are fishermen, but we are not dummies," McGill told the board. "The North Pacific is a rough place to make a living, and anything that enhances safety and seaworthiness of a vessel is an admirable goal."

In late 2007, Coast Guard marine inspectors traveled to the shipyard in Japan where the FCA had most of its dry dock work done. There was more work than could be finished in the scheduled time, and the *Ranger* left the yard with a number of work items unaccomplished. All the boats enrolled in ACSA were originally supposed to have met the program's safety requirements by January 1, 2008. By the time the *Alaska Ranger* sank three months later, the FCA ship still hadn't met many of the ACSA goals. Then again, neither had significantly more than half of the other head-and-gut boats.

It appeared that the Coast Guard's effort had fallen short. The Marine Board heard testimony that ACSA was underfunded and undermanned. The Coast Guard's Chief of Inspections for the state of Alaska revealed under questioning in Anchorage that he'd never even heard of ACSA until late in 2007, almost a year and a half after the program got off the ground. Clearly, there were communication problems between the two districts involved in bringing the boats up to the new standards. Inspectors were behind on their paperwork, and the Coast Guard's computer data system wasn't updated to effectively keep track of the ACSA work lists. But the Marine Board's fact-finding revealed that the underlying problem was that this fleet of boats was in much worse shape than the Coast Guard examiners knew when they started out. The ships' work lists were long, and the extensive dry dock time required to complete the repairs was hard to come by. It was true that most of the enrolled head-and-gut boats still hadn't met the ACSA standards. But almost all of them were in much better shape than they had been a few years before—or than they ever would have been if the local fishing vessel examiners hadn't pushed for the program in the first place.

WHILE THE COAST GUARD WAS FOCUSING on what had gone wrong with its alternative safety program, the National Institute for Occupational Safety and Health (NIOSH) in Anchorage was studying what had gone right. Fishing vessel safety expert Jennifer Lincoln had witnessed the ACSA program from its inception. She remembered the day when she'd driven to downtown Anchorage to meet her friend Chris Woodley for lunch. On his desk were the data from the National Marine Fisheries Service that proved the head-and-gut boats were illegally processing ancillary products. Woodley told Lincoln about his plan for an alternative safety program that would improve the seaworthiness of the Bering Sea head-and-gut fleet and showed her a list of additional safety requirements he and Charlie Medlicott had come up with based on the *Arctic Rose* and *Galaxy* casualty investigations. Woodley asked Lincoln what she thought.

She was thrilled. Woodley's program would require the head-and-gut boats to install life rafts on the rail of the ship that could be launched by a single person. There would be stronger requirements for training: at least five drill instructors on each boat that sailed with thirty-six or more crew. Study of earlier disasters had found that reflective tape and standard flashlights were sometimes not enough to allow rescuers to find survivors in the water—especially at night. As part of ACSA, all the survival suits on the Bering Sea head-and-gut fleet would be equipped with strobe lights.

On Sunday, March 23, 2008, Lincoln got a text message from Charlie Medlicott, the Coast Guard's fishing vessel examiner in Dutch Harbor. "*Alaska Ranger* sank. Forty-seven people on board," it read.

Lincoln called him. It wasn't just about his job, and the devastating fact that one of the ACSA boats sank just as the program they'd pushed so hard for was getting off the ground. Charlie knew the guys on that boat.

On Monday morning, Lincoln was on a plane to Dutch. She'd guessed that the *Alaska Ranger* sinking would warrant a Marine Board, and she'd read enough Coast Guard accident reports to know the board would focus on a wide array of topics. She wanted to be sure that data on training and survival equipment were well captured. It would be valuable for evaluating the changes that had been implemented as a result of ACSA. Lincoln knew that it was essential to gather the information immediately, while the survivors' memories were fresh and before they left town. Together with the NTSB investigators, Lincoln interviewed each of the fishermen before the formal questioning began. "What's your first language?" she asked each of the survivors. "Tell me about your immersion suit." "Did you have a strobe light?"

Lincoln quickly reached some first conclusions. It was apparent that the newly required strobe lights had saved lives. Until the very end of the rescue, there had been little "search" in this search and rescue case. The flashing lights on the fishermen's survival suits had directed the Coasties right to their targets. And though the abandon ship procedures had been anything but smooth, the life rafts were each successfully launched by one or two people. When the rafts bolted away from the boat, a number of individuals entered the water by climbing down the ship's Jacob's Ladders, survival equipment that was mandated through ACSA.

Lincoln found that 30 percent of the people on board—fourteen in all—had had recent safety training. The number wasn't ideal, but chances are it would have been lower without ACSA. Most significant, she discovered that 80 percent of the people who were recently trained got into a raft, while only 38 percent made it into a life raft without training. All of those who made it to a raft survived.

ON SEPTEMBER 30, 2009, A YEAR and a half after the *Alaska Ranger* was lost in the Bering Sea, NTSB investigator Liam LaRue presented his group's findings to President Barack Obama's recently appointed head of the National Transportation Safety Board, Deborah Hersman. As LaRue and his coinvestigators gathered their papers and settled in front of microphones, Captain Mike Rand and two of the other Coast Guard Marine Board members found seats together in the spacious auditorium. Their own report was still working its way up the Coast Guard chain of command, and the Coasties had yet to see the NTSB's findings. Most of the seats in the large room were empty, but a handful of people sat in the front row. Among them was Karen Jacobsen, Captain Pete Jacobsen's daughter.

"We are here today to make sure an accident like this does not happen again. That's the purpose of our work," began Hersman.

She turned to LaRue, who described the ship, its fishing practices, and its crew, then gave the board a condensed timeline of its last hours: The flooding was discovered in the *Alaska Ranger*'s rudder room around 2:30 A.M. There was no evidence of any collision or breach in the skin of the ship. LaRue's team had concluded that, as originally reported, the flooding had most likely resulted from the physical loss of one of the ship's two rudders. The board members had found that the *Ranger*'s conversion to a factory fishing trawler years before the accident had left the boat sitting more than two feet lower in the water than as originally designed. The difference was enough to bring the waterline to the top of the ship's rudder shaft and to allow a constant flow of seawater into the ship. Rough seas most likely made the situation worse.

Losing a rudder is a bad—and unusual—scenario. But it should not have caused the boat to sink. The NTSB's marine engineer explained to Hersman and her colleagues that if the *Ranger*'s watertight integrity had been intact, even complete flooding of the rudder room would not have led to the loss of the vessel. But the ship's watertight integrity was not, in fact, intact. Testimony had revealed that there was a permanent breach in the rudder room bulkhead, and that at least one of the ship's watertight doors had begun leaking not long after it was closed.

The final blow to the *Ranger* came when its controllable pitch propeller (CPP) system kicked into reverse soon after the vessel lost power. The CPP system was a remnant of the *Ranger*'s earlier life as an oil-rig supply boat. For a time, many ships were designed to automatically shift into reverse at an unexpected loss of power, the idea being that it would be far better to suddenly back away from an oil rig than ram right into it.

The NTSB concluded that the *Ranger*'s movement astern both

accelerated the sinking and prevented the fishermen from entering life rafts directly from the boat. The officers on board the ship did have the ability to shut down the ship's engines and stop the vessel's backward movement. There was no testimony or evidence, however, that any member of the crew attempted to do so.

In July 2008, three months after the accident, the Coast Guard had issued an industry-wide safety warning titled "Controllable Pitch Propeller Systems and Situational Awareness." There had been other recent incidents of CPP systems surprising crews with a shift into reverse. For instance, just a few months before the loss of the *Ranger*, a cruise ship carrying adventure tourists sank off the coast of Antarctica (all 154 passengers safely evacuated the ship before it shifted astern). The Coast Guard's warning encouraged owners and operators of boats with CPP propulsion to be sure they understood their system and knew how to react in the case of a sudden casualty. LaRue explained to Hersman that it was unclear if the officers on the *Ranger* actually understood the ship's CPP system. What was clear was that the ship and its crew would have fared better if the boat was dead in the water during abandon ship procedures, rather than moving full-speed astern.

Still, LaRue resisted judging the actions of men who weren't there to explain themselves. All of the *Ranger*'s top officers, after all, had died in the sinking: "We don't know what they knew and when they knew it, so it's hard to make a judgment on what they should have done," he told Hersman.

In the end, the NTSB's investigation resulted in five recommendations to four different groups. The Fishing Company of Alaska was advised to "review and modify as necessary the procedures for enforcing your drug and alcohol policy to ensure full crew compliance." Though there was significant evidence of substance use on the vessel, LaRue stated that the NTSB had been unable to determine whether alcohol or drugs had played a role in the incident since no drug or alcohol testing was conducted on the survivors.

"They definitely talked the talk but didn't follow through," LaRue responded to Hersman, after she quoted from the FCA's strongly worded no-tolerance drug and alcohol policy. "It appears that the policy was not enforced."

Though the NTSB had investigated the role of the fish master on board the vessel, LaRue said, they'd concluded that there was no compelling evidence that the fish master's power exceeded that of the captain, and had taken no action on the issue. Under questioning, former Captain Steve Slotvig had maintained that he left the *Ranger* of his own accord, and that he was, in fact, in control of the vessel.

The board's second and third recommendations were to the National Marine Fisheries Service and the North Pacific Fishery Management Council. Both organizations were told to amend their regulations to allow ships like the *Alaska Ranger* to be replaced in situations other than vessel loss without a company sacrificing its quotas.

The final two recommendations were directed at the Coast Guard. The Marine Board's investigations had revealed that the engineers on board the *Ranger* were not properly licensed. Chief Engineer Dan Cook held a Coast Guard license for vessels up to 6,000 horsepower. The *Ranger*'s horsepower was 7,000. Only one of the ship's two assistant engineers was licensed, and his license was for up to 4,000 horsepower. There was no reason to think, the investigators said, that the licensing of the engineers contributed to the casualty, but nevertheless their lack of qualifications was a notable oversight on the part of the FCA—and the Coast Guard.

The most significant of the NTSB's recommendations was one that the safety agency had made four times in the past twenty years, with no results. LaRue's group concluded that the Coast

Guard should go to Congress and seek the legislative authority required to regulate commercial fishing boats to an appropriate level of safety. It was clear that the *Ranger*'s demise was the result of a series of preventable malfunctions. Had the ship been required to adhere to higher standards of seaworthiness, the *Ranger* and her entire crew would very likely have made it safely back to Dutch Harbor.

The NTSB recognized that the problems the Coast Guard faced in improving safety in the commercial fishing fleet went far beyond the lack of dedicated resources for Woodley's ACSA program. The Coast Guard needed Congressional authority to regulate the entire industry to a common standard. Like every other type of boat, commercial fishing vessels should be inspected.

"There are a lot more regulations for the fish than there are for the fishermen's safety," Hersman noted in her concluding statements. "That needs to get rectified. . . . The *Deadliest Catch* is not called that for no reason. The statistics bear it out. This is the deadliest industry. . . . Whether you're on the *Cornelia Marie* or the *Alaska Ranger,* you should be assured of one level of safety.

"Issuing these recommendations is not the end. It's just the beginning of the process," she promised. "We're hoping the fourth time will be the charm."

Meanwhile, in Alaska and throughout the United States, fishing boats leave port every day, many of them headed to some of the most violent and unforgiving waters on Earth. As always, many of the vessels are crewed by greenhorns, with little or no experience in using the safety equipment that the owners of their boats are finally required to provide. When disaster strikes, they may be able to make their way into a survival suit or a life raft. As for the actual seaworthiness of their ship? As long as commercial fishing vessels retain their uninspected status, all the fishermen can do is pray.

Epilogue

Julio Morales hired a lawyer and agreed to a settlement that allowed him to buy a modest home in Southern California. He hasn't been working since the sinking, but has been thinking about getting another job in Alaska, maybe in the oil industry. Julio has some arthritis that he attributes to the accident, as well as quite a few nightmares. Every time he smells diesel fuel, he tastes it as well.

For almost a year after the sinking, several members of Julio's family refused to speak to him. They blamed him for Byron's death—for bringing Byron up to Alaska in the first place. Julio rarely sees his cousin Marco, who went right back to fishing. Marco worked for the FCA for another year, and then was arrested on an outstanding warrant as he tried to cross back into the United States after a trip to Mexico. He's still in jail.

Jeremy Freitag also got a lawyer, and ended up with enough cash to buy that house in central Oregon. He promised his mother he'd never go back to Alaska. He's studying for his captain's license and hopes to eventually run a sport fishing boat out of the tourist town of Newport, Oregon.

Eric Haynes is at home in Las Vegas, working toward a college degree in culinary management. He's thought about going back up to Alaska. He misses it sometimes, though he's not sure he'd be welcome back on another FCA boat after filing a lawsuit against the company.

A little more than a year after the sinking, Eric met Karen Jacobsen, Captain Pete's daughter, at the Fishermen's Memorial in Seattle. Karen had planned a sunrise service at the monument where her father's name—along with those of Byron Carrillo, Dan Cook, and David Silveira—had been inscribed months before. Karen was shattered by her father's death and sees great significance in the fact that the tragedy occurred on Easter—enough so that she chose to hold the ceremony at sunrise on Easter morning, rather than on the anniversary of the date the *Ranger* was lost.

In the two years since the boat sank, many more names have been added to the Seattle memorial, which is located at the working port known as Fishermen's Terminal. Several of the names are of men from the ninety-three-foot head-and-gut boat *Katmai,* a one-time shrimp trawler that sank out of Dutch Harbor six months after the *Ranger* (and that was not enrolled in the Coast Guard's ACSA program). There were eleven men on board. Seven died. At least two of the deceased *Katmai* fishermen had previously worked on FCA boats.

The ship was even farther away from help, and it took the Coast Guard many hours to reach the scene. Jayhawk pilot Shawn Tripp was one of the men on the case. He had been predeployed to Cold Bay, for the fall red king crab season. The future of aircrew predeployment to St. Paul, meanwhile, is expected to continue despite the long-anticipated permanent closure of the Coast Guard's LORAN station in February 2010.

* * *

THE COAST GUARD RESCUE TEAMS that responded to the *Alaska Ranger* case piled up awards, especially rescue swimmers O'Brien Starr-Hollow and Abram Heller, who was awarded one of aviation's highest honors: the Distinguished Flying Cross. The rest of the men on the crews of the 60 and 65 helicopters were each awarded Coast Guard Air Medals.

Soon after the sinking, Brian McLaughlin made the effort to get in touch with the families of the deceased fishermen. He talked with Karen Jacobsen; and he and his wife, Amy, met with Pete Jacobsen's brother, Billy, a retired tugboat captain who lives outside of Seattle. Forty-two lives were saved on March 23, 2008, but some of the Coast Guard rescuers had a hard time coming to terms with the fact that they hadn't saved all of them. McLaughlin gave his Air Medal to Billy Jacobsen, who keeps it on his living room mantel.

McLaughlin has since been promoted a rank to Lieutenant Commander, and in 2009 moved with his family to Mobile, Alabama, where he's now an instructor for new Coast Guard pilots. The family once again made the cross-country trip a summer-long RV journey.

Ed Cook continued working for the FCA for about six months after his brother's death. And he continued making notes about safety concerns on his daily record sheets. In April 2008, less than a month after the tragedy that killed five men, Ed repeatedly noted that the *Warrior*'s deck crew was tying open watertight doors at sea. By early 2009, Ed had found a new job.

Beyond Ed's documented concerns, the FCA continued to experience legal and safety problems. In the fall of 2008, the company agreed to pay a $449,700 fine to settle a series of environmental violations, including mishandling prohibited species, fishing in a protected marine area, and multiple counts of harassment against NMFS observers assigned to its boats. Just a couple of weeks before the settlement was announced, a senior crew member on board the *Alaska Juris* had contacted law enforcement to report that he'd been assaulted by the ship's fish master. In June 2009 the FCA trawler *Alaska Victory* cracked a hull plate while transiting through ice. And in early July, the *Alaska Warrior*'s Japanese boatswain fell overboard after becoming tangled in fishing gear. He wasn't wearing a life jacket. Like *Ranger* fish master Satoshi Konno, the boatswain's body has still not been found.

Notes

This book is a work of nonfiction. No scenes or characters have been invented. No names have been changed. The dialogue that appears in quotation marks is based either on recordings from the rescue, or, more often, the recollections of one or both of the people involved in the conversation. Thoughts attributed to individuals are based upon what that person later told me he or she was thinking at the time.

Throughout the book, I relied heavily on Coast Guard records from the rescue. There were several hours of audio recorded of the back-and-forth between the Communications Station in Kodiak and the *Alaska Ranger.* I also gained access to detailed search and rescue logs from the case, as well as written statements from many of the rescuers that were produced soon after the disaster. The joint Coast Guard/National Transportation Safety Board (NTSB) Marine Board of Investigation that followed the accident questioned dozens of people, and I occasionally relied on the transcripts of those interviews, as well as on other background documentation collected by the Marine Board. However, the vast majority of the scenes in this book are re-created from my own interviews with the survivors and rescuers involved in the case.

In some places, there were discrepancies in people's memories, particularly about the order of events and the passage of time between events. In those cases, I have tried to sort out the best truth I could determine. The fact remains that *Deadliest Sea* is a re-creation of an extremely stressful and chaotic event. There were forty-seven people on board the sinking ship, and dozens more people who played an important role in the subsequent rescue. Each of those individuals has his or her own story, filled with many details that I have neglected to recount here.

Following is a chapter-by-chapter explanation of sources.

Prologue

All of the communication between Coast Guard watchstander David Seidl and the *Alaska Ranger* is from audio recordings kept by COMMSTA Kodiak. I transcribed these recordings myself, and later compared my own transcriptions against those prepared by NTSB investigator Liam LaRue.

David Seidl submitted to repeated questioning and indulged my request to re-create the drive from his home in Kodiak to his Communications Station workplace. David's supervisor, Adam Conners, operations officer Phillip Jordinelli and others at the Communications Station were also generous with answering many questions about the role of the facility.

Kodiak was the third most profitable U.S. fishing port in 2008, a rank that has remained consistent for several years. Each year, the National Oceanic and Atmospheric Administration (NOAA), which is part of the Department of Commerce, collects catch and profit data related to commercial fishing and makes that data available and sortable on its Web site at www.st.nmfs.noaa.gov/st1/commercial/index.html.

Air Station Captain Andrew Berghorn, Public Affairs Specialist Kurt Fredrickson, and countless others in Kodiak provided helpful background information about the Coast Guard presence on the island—and throughout Alaska. National Marine Fisheries Service administrator Rob Swanson, who worked for years as a fisheries observer, also offered local insight and arranged a tour of one of Kodiak's fish-processing plants.

Chapter One

In Dutch Harbor, Charlie Medlicott and Christina Craemer provided invaluable local insight, as well as tour guide services. Brian Dixon, Paul Wilkins, Michelle McNeill, and Pastor Daniel Wilcox also each generously spent time talking with me about the Dutch Harbor community and/or fishing industry.

The Dutch Harbor fishing statistics are from NOAA data. New Bedford, Massachusetts, beats Dutch Harbor in terms of the total value of its catch (due mostly to the high price paid for East Coast shellfish), though in terms of total poundage harvested annually, Dutch Harbor is by far the most productive of any U.S. port.

Life on the *Alaska Ranger* was described in great detail by Eric Haynes, Evan Holmes, Jeremy Freitag, David Hull, Ryan Shuck, Kenny Smith, Alex Olivarez, Paul Munoz, Richard Reimers, and Julio Morales, among others. Further description of fishing practices came from the instructors at the Observer Training Center in Anchorage. The book *Ocean Treasures: Commercial Fishing in Alaska* (Alaska Sea Grant, 2003) also provided a helpful cross-check to information supplied directly by the fishermen.

The scenes involving former *Alaska Ranger* Captain Steve Slotvig are based on observations by the crew members mentioned above, particularly Jeremy Freitag, Ryan Shuck, and Kenny Smith, and by Marine Board testimony, including Slotvig's own. Under questioning from the Marine Board, Slotvig denied that it was the fight about ice that led him to leave the ship and said he could not recall the incident when others saw fish master Satoshi Konno spit at him.

Former *Alaska Ranger* Captain Richard Canty, now a tugboat captain who lives in Maine, also shared his experiences with the boat and FCA company culture.

A series of newspaper articles written by *Seattle Times* investigative journalist Hal Bernton provides further context on the Japanese influence on FCA vessels. Those articles can be accessed at http://www.seattletimes.com.

Friends and family of both Pete Jacobsen and David Silveira, as well as Captain Scott Krey and a number of crew on the *Alaska Ranger,* all confirmed that neither man was happy to be assigned to the ship.

Chapter Two

The C-130's role in the *Alaska Ranger* case was recounted by the two plane pilots, Lieutenant Commander Matt Duben and Lieutenant Tommy Wallin, and by Navigator Charles Helms.

The progression of flooding and preparation to abandon ship is based on interviews with Evan Holmes, Eric Haynes, Julio Morales, and Jeremy Freitag, as well as Marine Board testimony from each of those men and from Chris Cossich and Rodney Lundy.

The descriptions of Pete Jacobsen's history are from interviews with crew members, as well as with his daughter Karen Jacobsen, his brother, Billy Jacobsen, his niece, Jennifer Jacobsen, and his late wife, Patricia Jacobsen, who passed away in early 2009.

Chapter Three

Coast Guard HH-60 Jayhawk pilots Lieutenant Commander Brian McLaughlin, Lieutenant Steve Bonn, Commander Shawn Tripp, and Lieutenant Commander Zach Koehler all patiently recalled their March 2008 deployment to St. Paul, as did flight mechanic Rob DeBolt and rescue swimmer O'Brien Starr-Hollow. In the winter of 2009, I visited the LORAN station and gleaned additional insight from the men on that deployment, among them Commander Robert Gaudet, Lieutenant Chris Carter, flight mechanic Keith Bastman, and rescue swimmer Alex Major. Commanding Officer Steven Pfister and Chief Jennifer Shafer were also helpful, as were so many more of the 2009 LORAN staff and Kodiak aircrew members.

Discovery Channel producers Rosie Sharkey and Kyle Wheeler provided additional insight about winter on St. Paul, as well as friendly chauffer service between the LORAN station and the village.

Lieutenant Commander Michael "Scott" Jackson tracked down some hard-to-find information about the history of Coast Guard predeployment to St. Paul, which began in 1997.

The statistics about the dangers of commercial fishing come from the National Institute of Occupational Safety and Health (NIOSH). More information is available at http://www.cdc.gov/niosh/topics/fishing/. The crab-specific statistics were pulled from "Improving Commercial Fishing Vessel Safety Through Collaboration," written by Chris Woodley, Jennifer Lincoln, and Charlie Medlicott and published in the spring 2009 edition of the Coast Guard Journal *Proceedings*.

Aquilina "Debbie" Lestenkof of the Ecosystem Conservation Office of the Tribal Government on St. Paul provided perspective on the history of the island and life there today. Marine mammal researcher Andrew Trites, of the University of British Columbia, offered further scientific insight about the population of northern fur seals in the Pribilofs.

Public affairs staff members Natalie Granger, Angela Hirsch, and Ryan White at Coast Guard headquarters in Washington, D.C., were helpful in tracking down general Coast Guard statistics and historical facts, including those about the Coast Guard's role during Hurricane Katrina and the number of Academy graduates now active in the Coast Guard.

Chapter Four

The spring 2008 crew on board the cutter *Munro*—including Captain Craig Lloyd, Executive Officer Mike Gatlin, Operations Boss Jimmy Terrell, Chief Luke Cutburth, Operations Specialist Erin Lopez, Corpsman Chuck Weiss, and Junior Officers Paul Windt and Dan Schrader—recounted the details of the *Munro*'s role in the *Alaska Ranger* rescue. In late February 2009 Captain Lloyd generously allowed me to ride along during the ship's Bering Sea patrol. Crew members, including Ops Boss Brad Anderson and Junior Officers Andrew Brown, Ellen Moloi, Francesca Hanna, Caitlin McCabe, and the relentlessly helpful Crystal Hudak, were patient in answering my many questions about the ship, its crew, and its mission.

Coast Guard fishing vessel examiners Chris Woodley, Charlie Medlicott, Ken Lawrenson, and Marty Teachout provided an overview of the Coast Guard's role in examining commercial fishing boats. Mike Rosecrans and Richard Hiscock provided expertise on the political and legislative history and directed me to a number of written sources, including Hiscock's 2000 paper "Fishing Vessel Safety in the United States: The Tragedy of Missed Opportunities." Also helpful were back copies of the Coast Guard's

Proceedings magazine; the report "Living to Fish, Dying to Fish," produced by the Coast Guard's Fishing Vessel Casualty Task Force in 1999; records from recent Congressional testimony about safety in the fishing industry; and Chris Woodley's 2000 graduate thesis, "Developing Regional Strategies in Fishing Vessel Safety: Integrating Fishing Vessel Safety and Fishery Resource Management."

Peggy Barry kindly recounted the painful details of her son's death, as well as her family's triumph in pushing for increased safety standards on board commercial fishing vessels. She also shared many years' worth of newspaper clippings. The Coast Guard's library file on the *Western Sea* sinking was a source of additional documentation, including the Coast Guard's investigative report on the casualty, Alaskan police reports related to the incident, and autopsy results on Peter Barry and *Western Sea* captain Jerald Bouchard.

The decline in fishing deaths since the implementation of the Commercial Fishing Industry Vessel Safety Act in 1991 is documented by NIOSH and in the comprehensive Coast Guard report "Analysis of Fishing Vessel Casualties: A Review of Lost Fishing Vessels and Crew Fatalities, 1992–2007."

Charlie Medlicott and Chris Woodley recounted their memories of the *Big Valley* sinking. Details of the boat's loading condition were taken from the Coast Guard's informal investigative report into the incident, which includes search and rescue records from the case, and from related documents prepared by the ship's marine architect and obtained through the Freedom of Information Act (FOIA).

Chapter Five

Leslie Hughes of the North Pacific Fishing Vessel Owners' Association (NPFVOA) provided the 85 percent statistic for the amount of Alaskan fish caught by Washington boats. That percentage was confirmed by the Seattle environmental group Natural Resource Consultants. NPFVOA also runs the safety training programs where FCA sent some of their crew members. Hughes helpfully described some of the historic safety problems in the fishing fleet and changes in safety attitudes over the years. Jerry Dzugan, director of the Alaska Marine Safety Education Association (AMSEA), provided additional perspective that helped to inform the safety descriptions in this chapter and throughout the book.

Coast Guard Commander Chris Woodley provided many of ACSA's founding documents. Further description of the program in its current form is available at http://homeport.uscg.mil/mycg/portal/ep/contentDetailView.do?BV_&contentType=EDITORIAL&contentId=99696.

The accounts of the *Arctic Rose* and *Galaxy* tragedies are based on interviews with Coast Guard fishing vessel inspectors and on the Coast Guard reports on the casualties. There was a Marine Board of Investigation for the *Arctic Rose*. That 134-page document is available at http://www.uscg.mil/hq/cg5/docs/boards/Arctic%20Rose%20.pdf. Coast Guard personnel, as well as several NMFS fisheries observers, described the *Arctic Rose* as a ship with a poor reputation. The book *58 Degrees North* by Hugo Kugiya (Bloomsbury, 2005) recounts the boat's disappearance and the Coast Guard's extended investigation.

The Coast Guard's similarly in-depth investigative report into the *Galaxy* fire can be accessed at http://marinecasualty.com/documents/gal.pdf.

There were a number of discrepancies in individuals' memories of the order of events during the *Alaska Ranger* abandon ship efforts, particularly about the ship's losing power, shifting into reverse, and listing to starboard. I believe the most likely order is this one (and the NTSB's report drew the same conclusion) though many people did not notice the shift into reverse until after the list, and others did not notice it at all. Several people, including David Hull, described a large wave that pulled the fishing net off the deck. Others thought it was simply the rising water that sucked the net off the ship.

Chapter Six

Gwen Rains, Jayson Vallee, Christina Craemer, and Beth Dubofsky all told me about their experiences on FCA boats. NOAA fishery biologist and observer program administrator Martin Loafflad smoothed the way for me to sit in on the December 2008 training for new observers in Anchorage, where Mike Vechter and Dennis Moore provided an excellent overview of Alaskan fisheries and the job of fisheries observer. Amanda Saxton, Peter Risse, Rob Swanson, and Brian Dixon provided additional insight into the observer training requirements and the lifestyle of the typical fisheries observer.

Chapter Seven

The actions of Indio Sol and Chris Cossich are based on their own Marine Board testimony, and on the recollections of other men on the boat. Neither Sol nor Cossich responded to requests for interviews. The Joshua Esa scene is written from the perspective of Eric Haynes. The scene with Evan Holmes and P. Ton is written from Evan's perspective.

The Coast Guard's investigative report on the *Alaska 1* sinking was obtained through FOIA. Head investigator Alan Blume, now retired from the Coast Guard, provided additional insight and clarification after I reviewed his report.

Charlie Medlicott was one of the inspectors on the Coast Guard's *Alaska Spirit* investigation. The NTSB also investigated that incident, and I drew details from the agency's report, which is at http://www.ntsb.gov/Publictn/1996/mar9601.pdf.

I relied on the expertise of hypothermia and cold-water survival experts Dr. Martin Nemiroff and Dr. Alan Steinman, as well as on the fifth edition of the textbook *Wilderness Medicine* (Mosby, 2007). Both the Coast Guard and the National Marine Fisheries Service use the 1-10-1 rule in their training, and I drew on written training materials from the two groups. The "skinfold thickness" study is described in more detail on page 170 of *Wilderness Medicine,* in a chapter coauthored by Dr. Steinman.

Coast Guard Public Affairs Specialist Sara Francis shared her own memories of the *Selendang Ayu* crash, and I referred to Coast Guard press releases from the time. Several Coast Guard pilots described their "dunker" training. Navy Commander Mike Prevost, Lieutenant Commander Ellis Gayles, and Lieutenant John Mahoney at Air Station Miramar in San Diego, California, answered questions about the dunker training and allowed me to sit in on the two-day refresher course that every aircrew member periodically attends.

Chapter Eight

Master Chief Dave Hoover, Master Chief Clay Hill, Chief John Hall, Chief Doug Lathrop, and Petty Officer Dustin Skarra provided expertise on the Coast Guard's rescue swimmer program. Swimmers Wil Milam, Alex Major, Chuck Falante, and the entire November 10–15, 2008, class at the Advanced Helicopter Rescue School in Astoria, Oregon, shared additional insight into the training and mentality required to be an AST. Two books written by former Coasties further helped to inform my understanding of the job of rescue swimmer: *Brotherhood of the Fin* by Gerald R. Hoover (Wheatmark, 2007) and *So Others May Live* by Martha J. LaGuardia-Kotite (The Lyons Press, 2006).

The radio exchange between the 60 Jayhawk and the C-130 recounted at the end of the chapter is from an audio recording that was made by *Alaska Warrior* Chief Engineer Ed Cook. Ed was in the *Warrior*'s wheelhouse at the time, where the crew was able to overhear the exchange between the two Coast Guard aircraft on the ship's radio.

Chapter Nine

Background information on the *Alaska Warrior* and the ship's role in the rescue are based on interviews with Scott Krey and Ed Cook, and on the Marine Board testimony of FCA officers Scott Krey, Raymond Falante, Albert Larson, and Bill McGill. The descriptions of the *Warrior*'s poor repair are exclusively from Ed Cook, who shared photographs and videos from his time on the ship.

The radio exchange between Jayhawk pilot Brian McLaughlin and *Warrior* Captain Scott Krey is also from the audio recording made by Ed Cook.

Chapter Ten

The crew of the 65 Dolphin, Lieutenant Commander TJ Schmitz, Lieutenant Greg Gedemer, flight mechanic Al Musgrave, and rescue swimmer Abram Heller, each described their deployment on the cutter *Munro,* as well as the play-by-play of the rescue. Each man wrote his own statement about the events of the rescue soon after the case. I relied on those documents, as well as subsequent interviews, to re-create Byron Carrillo's fall from the basket. The perspective of *Alaska Ranger* engineer Jim Madruga was also valuable. Like rescue swimmer Heller, Madruga saw Byron Carrillo rise, apparently safe inside the basket, and did not realize there was a problem until he was inside the helo himself.

To lose a survivor from a rescue basket during a hoist is an extremely unusual occurrence in the Coast Guard. I asked a number of veteran rescuers if they had ever heard of it happening before, and identified only one other incident, also in Alaska, in 1998. In that case, a rescue of crew from a sunken fishing boat called the *La Conte,* the weather was more treacherous than in the *Alaska Ranger* case, and the rescue swimmer was not deployed. The flight mechanic lowered the rescue basket to the water and two men attempted to climb in together. One of them never got fully inside and fell to his death as the basket reached the helicopter. Two books were written about this sinking and both recount the incident: *The Last Run* by Todd Lewan (HarperCollins, 2004), and *Coming Back Alive* by Spike Walker (St. Martins Press, 2001).

Chapter Eleven

In addition to the survivors' own accounts of their medical care on the *Munro,* Corpsmen Chuck Weiss and Blake Mitchell Castillo, and Petty Officer Kelly Stearns detailed the crew's preparations to receive survivors.

Chapter Twelve

In addition to my interviews with Eric Haynes and his Marine Board testimony, I had the benefit of Eric's own detailed and descriptive ten-page statement titled "Recollections of the Sinking of the *Alaska Ranger*."

Much of the dialogue in the *Warrior*'s wheelhouse, including the rescued crew members' explanations of what happened to the *Alaska Ranger*, is transcribed from Ed Cook's personal recording.

The actions of Samasoni Fa'aulu and Mark Hagerman are as recounted by Julio Morales, Abram Heller, and the rest of the Jayhawk crew. I was unable to contact either Fa'aulu or Hagerman, and neither was called to testify before the Marine Board.

Chapter Thirteen

The rescue swimmers did not attempt CPR on David Silveira after he was pulled into the Jayhawk because at the time they believed there were still additional people to rescue. For the previous couple of hours, they'd been pulling people out of the water one after the other, and the expectation was that that pace would continue. Technically, Coast Guard rescuers are not required to commence CPR if they are more than an hour from advanced care, or if the victim has obvious physical signs that he or she has been deceased for a long period of time. When brought into the helicopter, Silveira was exhibiting early signs of rigor mortis, evidence that it was too late to help him. In retrospect, however, both rescue swimmers felt some regret that they had not attempted CPR on the fisherman.

Jennifer Lubrani and Dr. Jeffrey Pellegrino of the American Red Cross provided additional perspective on CPR standards and methods.

Paul Webb, a civilian search and rescue expert in Juneau, demonstrated the Coast Guard's modeling program and explained the factors that play into long-term survival in Alaskan waters. Chief of Response Mike Inman, Chief of Incident

Management Todd Trimpert, and watch standers Mike Glinksi, Jeremy Dawkins, and Nate Johnson in Juneau recounted the role District Command played in coordinating the rescue, including the recognition of the miscount. Public Affairs Officer Eric Eggen helped to coordinate my visit. Admiral Gene Brooks provided appreciated insight into the unique scope and challenge of the *Alaska Ranger* rescue.

Ryan Shuck's girlfriend, Kami Ottmar, shared the e-mail she received at her apartment in Spokane, Washington, on the evening of March 23, 2008.

Chapter Fourteen

Coast Guard Public Affairs Specialist Sara Francis wrote a series of wonderfully clear and detailed press releases about the first days of the Marine Board hearings into the *Alaska Ranger* case. Meagan Krupa attended the Anchorage hearing on my behalf and took copious notes. I attended most of the Seattle hearings myself, and later acquired transcripts of all of the Marine Board's interviews.

Brian McLaughlin kindly agreed to allow me to reprint his e-mail to Jimmy Terrell, which was first forwarded to me by Captain Craig Lloyd.

Jennifer Lincoln provided a copy of a document titled "*Alaska Ranger* Investigation Survival Factors," which summarized the findings from her early interviews with the crew members.

Coast Guard Marine Architects Brian Thomas and Steven McGee explained the stability factors that led to the loss of the *Ranger* in much greater detail than is relayed in this book. Their detailed report, as well as numerous other documents relevant to the Marine Board's investigation can be accessed at http://www.ntsb.gov/Dockets/Marine/DCA08MM015/default.htm.

The NTSB's full eighty-three-page report on the disaster can be downloaded at http://www.ntsb.gov/Publictn/2009/MAR0905.pdf.

The Coast Guard issued two Marine Safety Alerts in reaction to the loss of the *Alaska Ranger.* The first, on "Maintaining Vessel Watertight Integrity" can be viewed at http://www.uscg.mil/d9/msuchicago/docs/Safety%20Alert%201%2008.pdf.

The second, "Controllable Pitch Propeller Systems and Situational Awareness," is at http://www.marineexchangesea.com/AlertsAndBulletins/03-08%20re%20Controllable%20Pitch%20Propeller%20Systems%20....pdf.

As of early March 2010, the Coast Guard had still not released its own Marine Board report on the incident.

Epilogue

For more background on Seattle's Fishermen's Memorial and a link to similar monuments, see: http://www.seattlefishermensmemorial.org/.

NOAA lawyer Susan Auer and NOAA criminal investigator Nathan Lagerwey provided context on the environmental and observer-related fines levied against FCA. NOAA's press release on the $449,700 fine is at http://www.noaanews.noaa.gov/stories2008/20081125_alaskafisheries.html.

The assault on board the *Juris*, the ice-related accident on the *Victory*, and the summer 2009 man overboard death on the *Warrior* are a matter of Coast Guard record.

Updated information and additional links are at www.deadliestsea.com.

Acknowledgments

In the early days of reporting this book I felt that there was great luck behind the fact that every one of the Coast Guard rescuers was remarkably open, articulate, and thoughtful about the dramatic events of March 23, 2008. As I continued my research, I discovered that these attributes are typical of the Coast Guard culture.

Between April 2008 and October 2009 I spent time with the Coast Guard in Seattle, Oregon, Washington, D.C., and throughout Alaska. Every place I visited, I was greeted with friendliness, candor, and an eagerness to share knowledge of the Coast Guard's philosophy and mission. I owe thanks to every Coastie mentioned in this book, though I'd like to single out Greg Gedemer, Todd Troup, Crystal Hudak, Al Musgrave, Sara Francis, and Brian and Amy McLaughlin for their repeated assistance and—in several cases—incredible personal hospitality.

Charlie Medlicott, Chris Woodley, and Jennifer Lincoln explained and reexplained issues related to safety trends in Alaska's fishing industry—and the remaining hazards that continue to make commercial fishing such a dangerous profession. Their dedication to improving safety in the fishing fleet is nothing less than heroic.

I am ever thankful to all the *Alaska Ranger* fishermen who recounted the details of their ordeal, as well as so many other stories of their lives. Special thanks are due to Ryan Shuck and David Hull, the first *Alaska Ranger* survivors I met, who introduced me to the world of head-and-gut fishing. Julio Morales spent many hours with me and shared memories of his cousin, Byron. Eric Haynes, Evan Holmes, Kenny Smith, Alex Olivarez, and Jeremy Freitag provided further perspective from their time working in Alaska.

In the process of researching this book, I also had the pleasure of getting to know Karen, Billy, and Jennifer Jacobsen, Ed and Cindy Cook, and Celeste Silveira, who recounted the proud seafaring history of each of their families.

Rachel Levin, Elizabeth Bryer, Kathy Weiss, Maraya Cornell, and Jennifer Alessi provided invaluable feedback on early drafts of these chapters.

Thanks also to Angela Hirsch, Melissa Lasher, Meredith Lloyd Rice, Dawn MacKeen, Steve Byers, and James Vlahos, who listened to me obsess about this story and offered trusted advice and support.

David Schulte meticulously transcribed dozens of hours of interviews, and offered more than a few insights along the way. Alison Kelman and Greer Schott supplied fact-checking help. Caroline Hirsch expertly assisted in pulling together the photographs.

Thanks to my fantastic agent Laurie Liss and her team at Sterling Lord. My editor at William Morrow, Henry Ferris, and his assistant, Danny Goldstein, offered many excellent suggestions that helped to make this a much, much better book.

I am endlessly grateful to James Meigs and his talented team at *Popular Mechanics,* most of all Jennifer Bogo, David Dunbar, and Allyson Torrisi. Their fascination and enthusiasm for the story told in the original magazine piece continued through the writing of this book, and both inspired and invigorated me every step of the way.

Finally, thank you to my ever-supportive family and most of all to my wonderful husband, Dan Koeppel, for his endless patience, wisdom, and love.